To Halt or Not to Halt? That Is the Question

To Halt or Not to Halt? That Is the Question

Cristian S Calude
University of Auckland, New Zealand

World Scientific

NEW JERSEY • LONDON • SINGAPORE • BEIJING • SHANGHAI • HONG KONG • TAIPEI • CHENNAI • TOKYO

Published by

World Scientific Publishing Co. Pte. Ltd.
5 Toh Tuck Link, Singapore 596224
USA office: 27 Warren Street, Suite 401-402, Hackensack, NJ 07601
UK office: 57 Shelton Street, Covent Garden, London WC2H 9HE

Library of Congress Cataloging-in-Publication Data
Names: Calude, Cristian, 1952– author.
Title: To halt or not to halt? That is the question /
Cristian S. Calude, University of Auckland, New Zealand.
Description: New Jersey : World Scientific, [2024] | Includes bibliographical references and index.
Identifiers: LCCN 2024000360 | ISBN 9789811232275 (hardcover) |
ISBN 9789811232282 (ebook for institutions) | ISBN 9789811232299 (ebook for individuals)
Subjects: LCSH: Machine theory. | Computational complexity. |
Machine theory--Mathematical models. | Algorithms.
Classification: LCC QA267 .C343 2024 | DDC 004--dc23/eng/20240201
LC record available at https://lccn.loc.gov/2024000360

British Library Cataloguing-in-Publication Data
A catalogue record for this book is available from the British Library.

For any available supplementary material, please visit
https://www.worldscientific.com/worldscibooks/10.1142/12159#t=suppl

To the

"reader, whose brain is not a Turing Machine"[1]

[1] G. Longo [279, p. 21].

Foreword

The twentieth century featured several profound conceptual revolutions. Early on, relativity theory, special and general, of course. But even more revolutionary, quantum mechanics, much discussed by philosophers, with Feynman sums over all histories and qubit entanglement, and in mathematics, a body of work organised around the seemingly innocuous "halting problem," also a philosophical heavyweight, but much lesser known.

Now Cristian Calude has set out, in this extremely attractive and varied volume, to tell the story of why the halting problem is so revolutionary and so significant. Dare I even say, loved.

He attractively arrays before us many different versions of the halting problem and many different areas of application, showing both the fertility and richness of the new conceptual the toolkit that has evolved, through the better part of a century of work by a small community, to comprehend the full implications of an apparently inoffensive idea.

The result is an extremely inviting halting problem travel guide, a beautiful tapestry of new ideas and concepts, that marks a dramatic before and after break in the foundations of mathematical thinking.

Journey with Calude through this varied terrain of spectacular insights.

In an age when super AI's menace, it is good to see what the unfettered human spirit can accomplish and be proud.

Gregory Chaitin
author of
Algorithmic Information Theory
(CUB Press, 1987)

Preface

Computers seem so pervasive and powerful that we may be inclined to believe that they will eventually solve all problems. The opposite is true. Computers, classical and unconventional, like quantum ones, are drastically limited: not only are there problems unsolvable by computers, but these problems outnumber by far (in a sense to be made precise in the book) the problems computers can solve. And, then once proved, the algorithmic unsolvability of a problem is *forever*.

What sort of problems computers cannot solve, shortly are (algorithmically) unsolvable? Are they esoteric, imagined only in the minds of mathematicians? Are they peripheral and of no concern to the practice of computing science? The answers are negative: even some everyday problems people have to solve are algorithmically unsolvable.

This book tells the story of the "Halting Problem", arguably the most (in)famous unsolvable problem: can an algorithm decide in finite time whether an arbitrary computer program will eventually stop?

At first reading, this seems a dull, petty question: why could halting be of any interest? After all, you run the program and wait till it stops. But what if the program continues after a reasonable time, a week, a year, a decade? Can we infer that it will never stop? A short reflection suggests that we cannot say that "the program will never stop" no matter how long the computation has run without stopping. And, indeed, the *negative* answer was obtained over 70 years ago.[1]

Did this raise your interest? No? Maybe I should add that the answers to the following questions are negative because of it:

a) Can mathematics be done by computers only, without needing more

[1]For the reader interested in history, Turing did not formulate the problem, nor did he prove its undecidability.

than a dozen mathematicians?

b) Can software testing and validating be done by computers only, without the need for more than a few computer scientists?

c) Can we write an anti-virus program which never needs any update?

d) Can we write a perfectly secure operating system?

e) Will AI solve all practical problems of humanity?

I dare say this is interesting and somewhat disturbing, but confined to the world of computers ... Does the Halting Problem, directly or indirectly, "touch" problems in other domains? Our story includes a variety of such problems too. Here are some examples: Do humans have free will, or do our brains automatically generate thoughts and decisions? Is complete disorder possible? Do spurious correlations affect science, economy and biology? Can the rules of chess[2] allow a never-ending game?

Some halt/non-halt questions are challenging to answer even if they are not technically algorithmically unsolvable by computers. For example, to halt/non-halt life support, a marriage on rocks, a long friendship, etc. Agonising on such a question could trigger a significant degree of stress, anxiety and unhappiness. While there are no universal solutions, there are strategies [214] which relate to moving forward or not – halting or not. Some procedures to cope with undecidability may suggest new ways to unravel these questions' mysteries.

We interpret the solution of the Halting Problem not only in the standard negative way (as an impossibility) but also in a positive way. An example is productivity. In *Getting Things Done: The Art of Stress-Free Productivity* (Penguin Books, 2015), D. Allen stresses that defining the "next action" is essential in any efficient productivity system. Many "actions" can be algorithmically completed. Still, not all. Here are a few examples: a) solving a complex mathematical problem, b) writing an award-winning fiction book, and c) developing a new business idea. Can AI successfully finish undecidable tasks, in particular, those enumerated above? AI can help to some extent but cannot by itself complete them. C. Newport[3] argued that "the ability to consistently complete undecidable tasks is increasingly valuable in our information economy". Why? Because the skills needed for solving non-trivial instances of undecidable problems require deep knowledge and experience, creative intuition, persistence, physical agility, and much more. Even more importantly, these skills cannot be automated or

---

[2]Including the rule of the draw which forces a finite ending.

[3]`www.calnewport.com/blog/2014/11/30/on-undecidable-tasks-or-how-alan-turing-can-help-you-earn-a-promotion/`.

easily outsourced.

Sometimes my students asked unconventional questions about the Halting Problem. Here are a few questions and answers: i) What is your preferred application of the undecidability of the Halting Problem? The link between the Halting Problem and Gödel's incompleteness. ii) What is the most unexpected application of the undecidability of the Halting Problem? None after you understand the Halting Problem. iii) What is the most expensive consequence of the undecidability of the Halting Problem? The impossibility of automating software testing for semantic correctness.

The book proposes a journey which informs and engages the reader,[4] combines informality with formal arguments and gives many classical and new results, examples and applications. It aims to stimulate the curiosity of a large audience – including mathematicians, computer scientists, physicists, biologists, social scientists, philosophers, and undergraduate and graduate students – interested in the limits of computing and various ways to cope with them.

Writing this book was harder than I thought and more rewarding than I ever imagined. None of this would have been possible without Elena's support, occasional frustration, and collaboration with many colleagues and friends.

I am deeply indebted to all my co-authors of the articles used in this book: A. A. Abbott, J. M. Agüero Trejo, E. H. Allen, E. Calude, K. F. Celine, J. Conder, D. Desfontaines, M. J. Dinneen, M. Dumitrescu, A. Fedorov, Z. Gao, N. J. Hay, S. Heidari, R. Hua, N. Huang, S. Jain, H. Jürgensen, B. Khoussainov, F. Kroon, A. Kulikov, S. Legg, J. P. Lewis, W. Li, G. Longo, S. Marcus, W. Merkle, F. W. Meyerstein C. Müller, G. Păun, R. Navarathna, N. Poznanović, M. S. Queen, A. Salomaa, C. Shu, J. Sifakis, L. Staiger, M. Stay, F. Stephan, D. Thompson, K. Svozil, I. Ţevy, M. Zimand and D. Vaida.

During the (too) long period of working on the book, I have benefitted from discussions and comments on various topics covered in the book with many of my co-authors, with J.-P. Delahaye, S. George, T. Murphy, S. Yang, N. Yanofsky, L. Velez, and the excellent assistance of the World Scientific team, notably the book editor L. F. Kwong (we did it again!), the LaTeX expert D. R. Babu and the production editor G. Lee – my warmest thanks.

A huge thank you to Y. Kolezhitskiy for carefully reading various drafts of the book and making pertinent suggestions for improvement.

[4]Check the footnotes with "Challenge for the reader".

Using the book's title, I "generated" the cover picture with a free AI Image Generator. I am grateful to P. Delmas and the World Scientific cover designer C. C. Ng, for enhancing it. I also acknowledge using the application Grammarly in some parts of the book.

I thank V. Chaitin for granting permission to use her picture of G. Chaitin, J.-P. Delahaye for granting permission to reproduce the table "Petite histoire de l'indécidablilité" from his article, Presque tout est indécidable, *Pour La Science* 375 (2009), 88-93, and G. Longo and Y. Matiyasevich for granting permission to use their pictures.

All pictures reproduced from Wikipedia comply with file's license requirements `commons.wikimedia.org/wiki/Commons:Reusing_content_outside_Wikimedia`.

Cristian S. Calude
The University of Auckland
New Zealand

November 2023

Contents

List of Figures

List of Tables

Chapter 1

From Entscheidungsproblem to the Halting Problem

In this chapter, we discuss the Halting Problem. We start with algorithms, pseudocode and examples of programs which halt or do not halt, then formulate the problem, look at decidable cases, prove the Halting Theorem (undecidability), and explore related problems and solutions.

1.1 Why the Halting Problem?

The Halting Problem is historically important because it was one of the first problems to be proven undecidable, i.e. not solvable by 'classical' computers, the type of device the reader has on the desk. Classical computers are physical implementations of the mathematical notion of deterministic Turing machine [151].[1]

The Halting Problem is a mathematical question regarding deterministic Turing machines.[2] According to Copeland [138, p. 40]

> The halting problem was so named (and, it appears, first stated) by Martin Davis [147, pp. 70–71]. The statement that the halting problem cannot be solved by a computing machine is known as the 'Halting Theorem'. (It is often said that Turing stated and proved the halting theorem in 'On Computable Numbers',[3] but strictly this is not true.)

[1]In the classical approach to computing, information is stored in bits that are represented logically by either a 0 (off) or a 1 (on). Classical computations use deterministic processors like x86 and ARM; repeating them with the same inputs produces the same output. In contrast, unconventional computers, like quantum, biological and other types of computers [70], use different means to store and process information.

[2]Turing invented the Turing machine in [397]. According to [270], Gödel developed the idea of a computing machine which closely resembles the Turing machine.

[3][397].

The result cited in [147, p. 70] is

> **Theorem 2.2.** *There exists a Turing machine whose halting problem is recursively unsolvable.*[4]

Fig. 1.1: M. Davis[5]

In his 1958 book, Davis credits Kleene's monograph [248, p. 382] with an informal proof. As Copeland noted, the paper [397] is incorrectly cited as the place where the Halting Theorem was first stated and proved. What is the reason that this erroneous statement is persistently re-asserted? We will answer this question in detail in Section 7.5.

1.2 A proof in verses

G. Pullum, from the University of Edinburgh, published the following enjoyable poetic proof of the Halting Problem [329]:[6]

> No general procedure for bug checks will do.
> Now, I won't just assert that, I'll prove it to you.
> I will prove that although you might work till you drop,
> you cannot tell if computation will stop.

[4]Recursively unsolvable is the old name for effectively/algorithmically unsolvable, undecidable or incomputable.

[5]Source: In Wikipedia. `en.wikipedia.org/wiki/Martin_Davis_(mathematician)#/media/File:Martin_Davis.jpg`.

[6]Originally published in *Mathematics Magazine* 73, 4, 319–320, in October 2000.

For imagine we have a procedure called P
that for specified input permits you to see
whether specified source code, with all of its faults,
defines a routine that eventually halts.

You feed in your program, with suitable data,
and P gets to work, and a little while later
(in finite compute time) correctly infers
whether infinite looping behavior occurs.

If there will be no looping, then P prints out 'Good.'
That means work on this input will halt, as it should.
But if it detects an unstoppable loop,
then P reports 'Bad!' which means you're in the soup.

Well, the truth is that P cannot possibly be,
because if you wrote it and gave it to me,
I could use it to set up a logical bind
that would shatter your reason and scramble your mind.

Here's the trick that I'll use and it's simple to do.
I'll define a procedure, which I will call Q,
that will use P's predictions of halting success
to stir up a terrible logical mess.

For a specified program, say A, one supplies,
the first step of this program called Q I devise
is to find out from P what's the right thing to say
of the looping behavior of A run on A.

If P's answer is 'Bad!', Q will suddenly stop.
But otherwise, Q will go back to the top,
and start off again, looping endlessly back,
till the universe dies and turns frozen and black.

And this program called Q wouldn't stay on the shelf;
I would ask it to forecast its run on itself.
When it reads its own source code, just what will it do?
What's the looping behavior of Q run on Q?

If P warns of infinite loops, Q will quit;
yet P is supposed to speak truly of it!
And if Q's going to quit, then P should say 'Good'.
Which makes Q start to loop! (P denied that it would.)

No matter how P might perform, Q will scoop it:
Q uses P's output to make P look stupid.
Whatever P says, it cannot predict Q:
P is right when it's wrong, and is false when it's true!

I've created a paradox, neat as can be
and simply by using your putative P.
When you posited P you stepped into a snare;
Your assumption has led you right into my lair.

So where can this argument possibly go?
I don't have to tell you; I'm sure you must know.
We've proved that there just cannot possibly be
a procedure that acts like the mythical P.

You can never find general mechanical means
for predicting the acts of computing machines;
it's something that cannot be done. So we users
must find our own bugs. Our computers are losers!

The reader, convinced by the proof above, could skip the rest of the chapter.

1.3 Pseudocode programs

An algorithm is a step-by-step set of operations that must be executed or performed. Algorithms can be executed in mind, on paper or by a computer.

Pseudocode is an informal high-level description language written in English-like syntax, which uses the structural conventions of standard programming languages but is intended for human reading rather than machine execution.

Pseudocode is not just for beginners; experienced programmers occasionally write pseudocode when collaborating with team-mates or even for themselves and use it in articles and textbooks. For example, pseudocode in data science or web development is used to describe the distinct steps of an algorithm in a way that is easy for anyone with basic programming knowledge to understand. Its intuitiveness helps the development and debugging.

No standard for pseudocode syntax exists; the level of detail of the pseudocode differs, and in some cases, it approaches that of formalised general-purpose programming language, i.e. a programming language equivalent in computational power with a universal Turing machine [50].

Pseudocode programs use typical instructions in standard programming languages like:

(1) assignments $(=)$,
(2) comparisons $(=, \neq, \leq, \geq, <, >)$,
(3) arithmetic operations $(+, -, \times, /)$,
(4) logical operations $(\vee, \wedge)$,
(5) programming keywords (**read**, **print**, **do**, **for**, **stop**),
(6) conditionals (**if-then-else**, **case**), iteration (**while**),
(7) wrapping (**begin**, **end**), and
(8) functions $(\mathbf{P}(N))$.

For example, $i = 1$ denotes the assignment of the value 1 to the variable i; $s = s+1$ is the instruction that increments the current value of the variable s with 1; **while** $i \leq 100$ **do** is a control flow statement that repeats the execution of a block of instructions as long the Boolean condition $i \leq 100$ is satisfied.

We will require *each program ends with the unique instruction* **stop**.

1.4 Halting programs

The (pseudocode) program

```
PROGRAM(1)
i = 1
s = 0
while i ≤ 100 do
      s = s + i
      i = i + 1
end while
print s
stop
```

calculates the sum $1+2+\cdots+100$: it outputs the number 5,050 and **halts.**[7] Deciding whether this program halts is easy: the condition "$i \leq 100$" in the iteration **while** (lines 3, 4, 5, 6) is satisfied only 100 times, hence on the 101-th iteration, the condition fails. The control goes to instructions 7 and 8, which print 5,050 and terminate the computation. To facilitate the understanding of programs, instructions will be numbered. These numbers are not part of the actual program; they are meta-information.

1.5 Non-halting programs

How could a program run forever? Figure 1.2 suggests such a behaviour.

Fig. 1.2: You first!

An example of such a program can be obtained by modifying PROGRAM(1):

[7]The eight-year-old Gauss, who went to become the *Princeps mathematicorum*, amazed his teacher when he quickly calculated this sum. Gauss didn't use a computer but noticed that the sum consists of fifty pairs of numbers $(1+100), (2+99), (3+98), \ldots, (50, 51)$ of constant value 101, that is $50 \times 51 = 5,050$. The more you think, the less you compute and vice versa!

```
PROGRAM(2)
1 i = 1
2 while i ≥ 1 do
3       i = i + 1
4       print i
5 end while
6 stop
```

Deciding whether PROGRAM(2) halts is also easy. The condition "$i \geq 1$" in **while** (lines 2, 3, 4, and 5) is always satisfied; hence the instructions 2, 3, and 4 are repeated again and again, forming a loop: PROGRAM(2) never halts, and the instruction **stop** is never reached. This program prints all positive integers, $1, 2, 3, \ldots$

1.6 Loops and non-halting programs

For more insight, let us look more closely at PROGRAM(2): this program never stops because of the infinite repetition of the instructions on lines 2, 3, and 4. The execution of the program can be described by the following *infinite* sequence of instructions:

$$1, 2, 3, 4, 2, 3, 4, 2, 3, 4, \ldots 2, 3, 4, \ldots$$

In other words, PROGRAM(2) never halts because it executes the *loop* $2, 3, 4$ forever, so the instruction on line 5 is never reached. In mathematics, a sequence $u_1, u_2, \ldots, u_n \ldots$ is called *periodic* if it repeats itself over and over again at regular intervals. Formally, the sequence is periodic with period $T > 0$ if $u_{n+T} = u_n$, for all $n \geq 1$. The smallest such T is called the sequence's least period (or just "the period"). For example, the sequence $0, 1, 0, 1, 0, 1, \ldots$ is periodic with period 2, and the sequence $0, 1, 1, 1, 0, 1, 1, 1, \ldots$ is periodic with period 4.

In computer science, an *infinite loop* is an indefinitely repeated set of instructions. For example, a program that executes the instructions 1,2,3,4,5,6 in the following order

$$1, 2, 3, 4, 5, 6, 3, 4, 3, 4, 3, 4, \ldots$$

has a loop $(3, 4)$ and, consequently, it never stops. This sequence is periodic. A program generally contains an infinite loop if the program executes a periodic sequence of instructions on an input. However, in contrast to

PROGRAM(2), it is possible to write a program with an infinite loop which stops on some inputs.[8]

A program whose sequence of instructions contains an infinite loop never halts, and deciding if a program contains a loop is easy.[9] What about the converse implication? In other words, is it true that any computation which never stops has to contain an infinite loop?

The answer is *negative*, and PROGRAM(3) is an example.

```
PROGRAM(3)
1  i = 1
2  while i ≥ 1 do
3      j = 1
4      while j ≤ i do
5          print 0
6          j = j + 1
7      end while
8      i = i + 1
9  end while
10 stop
```

Indeed, the sequence of instructions executed by PROGRAM(3) starts with

$$1, 2, 3, 4, 5, 6, 7, 8, 9, 2, 3, 4, 5, 6, 7, 4, 5, 6, 7, 8, 9, \ldots$$

and never ends. However, the sequence does not have an infinite loop because the number of repetitions of the sequence $4, 5, 6, 7$ grows indefinitely.

A *loop instruction* is a control flow instruction that allows a block of instructions to be executed repeatedly. For example, **while** is such an instruction for the programs used in this book. Some loops are finite: the loop in PROGRAM(1) is finite. The infinite loop realised by the group of instructions $2, 3, 4, 5, 6, 7, 8, 9$ shows that PROGRAM(3) never halts. Another example is a program in which two instructions, a and b, are executed in the following sequence:

$$a, b, aa, ab, ba, bb, aaa, aab, aba, abb, baa, bab, bba, bbb, aaaa, \ldots \tag{1.1}$$

This never-ending binary sequence contains every binary string infinitely many times.[10] One can write a program with the instructions labelled a and

[8] Challenge for the reader: Write such a program.
[9] Challenge for the reader: Why?
[10] Challenge for the reader: Why?

b whose execution consists of exactly the sequence (1.1): this program will not halt, but it does not have an infinite loop.[11] These types of programs are not only non-halting but form the "majority" of non-halting programs.[12]

1.7 The Halting Problem

As we have seen in Sections 1.4 and 1.5, deciding that PROGRAM(1) always halts and PROGRAM(2) never halts was easy[13], so perhaps with a bit more thinking we can find a solution which works for every program. But what about the discussion in Section 1.6?

> The *Halting Problem* (for programs) is the problem of deciding whether an arbitrary program halts.

To solve the Halting Problem (for programs), we need a program called **halting decider**, which can correctly predict in finite time whether or not any other *program* stops.[14] There are no limitations on the amount of memory or time required by the **halting decider** to make the prediction. However, the **halting decider** should return the decision in *finite time*; hence **halting decider** always stops. As programs can include their inputs, the **halting decider** should be able to handle also a program and its input.

Confirming that a stopping program eventually halts can be done just by running the program, as there is no time limit for reaching the decision. What about the case when the program does not stop? How long should we run the program to decide it will never stop? Intuitively, this time could be very (arbitrarily?) large. But, is it always *finite*?

1.8 Infinitely many decidable cases

The Halting Problem is decidable for various infinite classes of programs and inputs. For example, every program which contains no loop instruction halts, and there are infinitely many such programs.

We can construct many other examples of decidable classes, and here are examples of two such classes. Start with PROGRAM(1) and "parametrise" **while** (replace the bound 100 by the variable N) to obtain the program

[11]Challenge for the reader: Write such a program.

[12]Challenge for the reader: Formalise the notion of "majority" and justify the above statement.

[13]Note that both programs don't have inputs.

[14]The future tense indicates that the decision has to be made 'in advance'.

PROGRAM(4)

```
read N
i = 1
s = 1
while i ≤ N do
      s = i + s
      i = i + 1
end while
print s
stop
```

which clearly halts and calculates the sum $1 + 2 + \cdots + N = N(N+1)/2$. If we vary the input value $N = 1, 2, 3, \ldots$, we get infinitely many (similar) halting programs. Analogously, starting from PROGRAM(2), we obtain infinitely many (similar) never-halting programs:

PROGRAM(5)

```
read N
i = N
while i ≥ N do
      i = i + 1
      print i
end while
stop
```

Can we now conclude that the Halting Problem is decidable? Recall that to prove decidability, we need to show that there is a **halting decider** which correctly solves every instance of the Halting Problem in finite time. Deciding correctly in finite time infinitely many cases is only a step toward a positive answer, not the answer.

The examples above are not too "clever": however, they are essential because of the possibility to replicate them in infinitely many slightly different ways. Next, we discuss another method to obtain decidability based on additional information. To this aim, we introduce a vital program parameter, the *length.* The motivation is the following. Programs are written using a finite alphabet. For example, the C programming language uses 256 characters. *The number of characters a program uses is its length.* For example, PROGRAM(1) uses 46 characters, including a space after each instruction (we do not count the meta-characters for line numbers and the program's name).

Decidability Criterion. *Fix a program P and $n > 0$. If we know the number N of elements in the set*

$$S = \{(P, v) \mid P(v)\,\text{halts and length}(v) \leq n\},$$ [15]

then the Halting Problem for S is decidable.

Here is the proof. First, the number of programs with length at most n, where n is a fixed positive integer, is finite because the number of strings of bounded length over a finite alphabet is finite.[16]

Second, if we know the number N of the set S elements, the following algorithm decides the Halting Problem for the set S. On input (P, v):

1) enumerate all strings v' with length$(v') \leq n$ and for each string, start running the computation $P(v')$;[17]
2) stop all computations when N halting pairs (P, v') have been obtained, and denote by S' their set;
3) if $(P, v) \in S'$ then $P(v)$ stops; otherwise, $P(v)$ does not halt.

The condition $(P, v) \in S'$ is decidable, and the algorithm is correct because S' has N elements.

The Decidability Criterion requires that we are given P, n and N. Obviously, we cannot replace N with the weaker condition "there exists an $N > 0$". More interestingly, we cannot replace n with "there exists an $n > 0$". The Decidability Criterion can be slightly generalised by replacing P with a finite set of programs $P_1, P_2, \ldots, P_j$, but *not* with an arbitrary computable set of programs.[18]

As the argument above depends on a parameter n, we again have an infinite class of decidable instances of the Halting Problem.

1.9 An infinite decider

The following program decides the halting of any pair $(\mathbf{P}, N)$:

[15] Recall that $P(v)$ denotes the computation of the program P on the input v.

[16] The number of programs P using M characters and length less or equal to n is at most $1 + M + M^2 + M^3 + \cdots + M^n = (M^{n+1} - 1)/(M - 1)$.

[17] Some computations will halt, some will not.

[18] Challenge for the reader: Can you prove the last claims?

```
PROGRAM(6)
read P, N
P(N)
print YES
stop
```

PROGRAM(6) reads an arbitrary pair (**P**, N) and executes **P**(N): if the computation halts, then PROGRAM(6) stops and returns YES, and otherwise, it continues forever.

Does this program solve the Halting Problem? Recall that a program solving the Halting Problem must *correctly and in finite time* decide whether it eventually halts for every pair (program, input). PROGRAM(6) correctly returns YES in case **P**(N) stops but continues *forever* on the contrary case, so it violates the finite time condition.

1.10 The Halting Theorem

We now prove the undecidability of the Halting Problem for pseudocode programs:

> **The Halting Theorem.** *There is no program which can solve correctly and in finite time the Halting Problem for every input pair.*

We prove by absurdity[19] and assume that there exists a **halting decider** solving the Halting Problem for all programs using positive integers for input/output, a strict sub-class of all programs. Suppose we show that this assumption leads to a contradiction. In that case, we have a proof of the undecidability of the Halting Problem.[20]

The proof below – proposed by G. Chaitin, see [61] – does not use Cantor's diagonalisation,[21] in contrast with the standard proofs, see for example [362, p. 207]. The idea is to assume the existence of a **halting decider** and use it to construct a simple self-contradictory program PROGRAM(7). The program uses a positive integer N as input and operates on the finite set of programs no longer than N bits in size. The contradiction

[19] A proof by absurdity, in Latin *reductio ad absurdum*, is an argument establishing a claim by showing that its negation would lead to absurdity or contradiction.

[20] Restricting the class of programs does not diminish the generality of the proof. Challenge for the reader: Why?

[21] We illustrate this type of proof in Section 5.6.

is caused by PROGRAM(7) generating itself if N is large enough because its length becomes smaller than N, giving contradictory outputs.

Here is the program:

```
PROGRAM(7)
1 read N
2 generate all programs P with length(P) ≤ N
3 run halting decider on all programs above
4 delete all non-halting programs
5 run the remaining halting programs till all stop
6 compute the biggest output w produced by the halting programs
7 assign v = w + 1
8 print v
9 stop
```

PROGRAM(7) uses not only the instructions in Section 1.3 – like the first five programs – but also programs such as **generate**, **run** or **remove**. However, these programs *can* be programmed using the instructions in Section 1.3. For example, the program **generate** enumerates all strings of length less than or equal to N and removes from the list those strings which are *not syntactically correct programs.*[22]

PROGRAM(7) reads a positive integer N (line 1), generates all programs P of length less than or equal to N, uses only positive integers as variables (line 2), and the condition length$(P) \leq N$ is satisfied only by finitely many programs.[23] Then it runs **halting decider** on each generated program (line 3)[24] and, based on these decisions, removes all non-halting programs (line 4). Then, the remaining programs, i.e. those which eventually halt, are run till they stop (line 5) – a fact guaranteed by the correctness of the **halting decider** – and assigns to w the maximum output returned by each of them (line 6). On line 7, $v = w + 1$; on line 8, v is printed, and the computation stops. Note that v is bigger than any value computed by every halting program of length at most N.

> PROGRAM(7) *stops* on each input N and *for large enough* N *the length of* PROGRAM(7) – which depends only on N – *is smaller than* N.[25]

[22] Challenge for the reader: Is the condition in Italics decidable? Justify your answer.

[23] As we noted in Section 1.8

[24] Recall that **halting decider** always stops and returns YES or NO.

[25] Challenge for the reader: Justify this claim.

We note that for large inputs N, PROGRAM(7)

(1) generates itself (line 2),
(2) it outputs a positive integer which is less than v, the maximum output produced by the halting programs of length less than N,
(3) it outputs v,

hence a contradiction. The crucial step in justifying (1) – (3) above is to calculate the length of PROGRAM(7): the program has a constant length (the sum of all instructions lengths) plus the length of the input read N, which is at most $\log_2 N + 1$ bits. *For large N, the length of* PROGRAM(7) *is smaller than N*, hence PROGRAM(7) **generates itself**, so it produces two different outputs, a contradiction.

The existence of the **halting decider** leads to the contradiction; hence the Halting Theorem was proven.

The Halting Theorem shows that *there is no single algorithm that can tell whether any program halts.* Here are two natural questions:[26]

(1) *Does the Halting Theorem* imply the existence of at least one program for which we cannot decide whether it halts or not?
(2) *Does the Halting Theorem* mean that there is any single program for which no algorithm can correctly decide whether or not it halts?

1.11 Can humans solve the Halting Problem?

The answer may seem affirmative. After all, a programmer can often look at a program and tell whether it will halt. But this ability *cannot work correctly for every program.* Indeed, to "solve" the Halting Problem, one has to be able to study any program and decide correctly in a finite amount of time whether it halts or not: *it is not enough to succeed on some programs, or even on a finite very large set of programs.*[27]

Programs written by humans are usually straightforward and are designed to terminate. Given enough times, can humans decide the termination of humanly written programs?

Programmers can fail in making correct decisions because of reasons different from their inability to solve infinitely many instances of the Halting Problem. Can humans always tell whether even short programs halt? For

[26] Challenge for the reader: Answer the questions (1) and (2) and give justifications.
[27] Humans can deal only with finitely many cases.

example, a programmer may oscillate between the decisions halt and non-halt, at one moment preferring one and at another moment opting for the other one, depending on many factors. Why? Because according to [392], human decision-making is not just a neutral calculation or step-by-step logical reasoning, but "rather a parallel process carried out by a neural network that simultaneously tries to evaluate the attractiveness of various options". Indeed, a programmer's decision may oscillate because focusing on the program's different properties could lead to "illusory correlations", relationships that do not exist. The reverse situation can happen too: even if the programmer makes the correct decision, say, the program stops, the decision could not be practically relevant because the time till the program stops is exceedingly long.

Either with humans or with computers – there isn't that much difference.

Can a hybrid approach involving a testing program and a human analysis solve the Halting Problem? Again, for the reasons described above, the answer is negative. But what about solving the Halting Problem for a large finite set of programs of practical interest? After all, in computing science, only such programs matter. There is no general result,[28] but we will briefly discuss an example in Section 2.8.

1.12 Church-Turing Thesis

An *algorithm*[29] is a finite and completely specified initial-step-to-concluding-step procedure for solving problems. Every step is readily executable by an agent with limited cognitive resources. The rules for moving from one step to the next are entirely determinate and self-contained, and its output is delivered, if at all, after a finite number of steps.

A function $f : X \to Y$ is called *effectively/algorithmically computable* if there exists an algorithm that an *agent*[30] can mindlessly execute to compute for every $x \in X$ the value of $f(x) \in Y$. The computation is executed by an algorithm, hence it stops in a *finite* time, a short or huge one.

[28]Challenge for the reader: What is your opinion?

[29]The word 'algorithm' comes from the name of Persian mathematician Muhammad ibn Musa al-Khwarizmi.

[30]A "computist" for Turing [397] and a "worker" for Post [324].

Fig. 1.3: M. ibn Musa al-Khwarizmi[31]

Fig. 1.4: K. Gödel[32]

Using the coding devised by Gödel in his seminal article [194] (called *arithmetisation* or *gödelisation*), one can restrict without loss of generality

[31]Source: In Wikipedia. `commons.wikimedia.org/wiki/Category:Muhammad_ibn_Musa_al-Khwarizmi#/media/File:Muammad_ibn_Ms_al-Khwrizm.png`.

[32]Source: In Wikipedia. `en.wikipedia.org/wiki/Kurt_Gödel#/media/File:Kurt_gödel.jpg`.

this discussion to functions from natural numbers, possible with multiple arguments, to natural numbers. For example, a function from programs to graphs can be completely represented by such a function. The notion of effectively/algorithmically computable function is *informal* as it is based on the non-mathematical (vague, for some) concepts of algorithm and agent.[33]

In the 1930s, four formal definitions for the effectively computable functions were proposed, namely:

(1) *Gödel general recursive functions* in 1931 [194], with partial input from an unpublished suggestion of Herbrand [248, p. 274],
(2) *Church λ-definable functions* in 1936 [126],
(3) *Post computable functions* in 1936 [324] and
(4) *Turing computable functions* in 1939 [398].

Fig. 1.5: A. Church[34]

Amazingly, these different mathematical definitions are *equivalent* as proven in [126][35] and [398]. Based on these equivalences and heuristic

[33]An interesting critique of the concept of "vagueness" in this context is discussed in [366].
[34]Source: In Wikipedia. `en.wikipedia.org/wiki/Alonzo_Church#/media/File:Alonzo_Church.jpg`.
[35]Independently obtained by Church and his student Kleene.

evidence, a "thesis" stating that each of these classes faithfully describes formally the informal class of effectively computable functions has emerged.

The first form of this "thesis" was stated by Kleene [248]:

> **Thesis I.** Every effectively calculable function is general recursive.

Equivalently,

> **Church-Turing Thesis.** Every effectively calculable function is Turing computable.

Half of the Church-Turing Thesis, if a function is (general) recursive or Turing computable, it is undoubtedly effectively computable, is relatively unexciting. The other half is interesting (and puzzling, even after almost 70 years since it was stated).

Church-Turing Thesis can be used to simplify proofs. Instead of writing a complete description of a Turing machine and proving that it calculates a specific function, we write an informal algorithm and argue that it does the required job. In particular, pseudocode programs used in this book are computationally equivalent to Turing machines so that we can substitute programs with Turing machines in any statement about programs. Of course, formal proofs of different other classes of functions being equal to the class of general recursive functions, including classes of functions computed by different types of programs, have been given; see [50].

Furthermore, Church-Turing Thesis gives interesting interpretations to technical results: prime examples are the negative answers to Entscheidungsproblem and Hilbert's Tenth Problem [293].

Church-Turing Thesis *is not* an issue about the computational limits of all possible machines (whatever they may be) – for a much more interesting question, see hypercomputation [24] – but about the adequacy of the Gödel/Church/Turing equivalent models for the class of effectively computable functions. From this perspective, no serious challenge to the thesis has been presented so far despite many attempts [243, 147, 256]; see also [311].

Can Church-Turing Thesis be proven, or is it instead essentially just a definition of computability? As its name indicates, the statement is a "thesis" and not a theorem because it states an equivalence between a non-mathematical notion and a mathematical one. The statement's validity can be argued but cannot be (mathematically) proven; however, it could be "weakly" disproved by producing a solid example of an effectively calculable function that a Turing machine cannot compute. This argument is accepted almost unanimously [140]. More recently, in [296, 366], inconclusive arguments in favour of provability have been proposed[36], while unprovability was re-defended in [52]. An axiomatisation of classical algorithms has been proposed in [158], and a proof of the Church-Turing Thesis was given in this framework.

Fig. 1.6: A. Turing[37]

An interesting argument which goes back to Gödel [177] has been revisited in the context of recent advances in AI (see [284, 124]):

> Turing in his 1937 [cited paper[38]] gives an argument which is supposed to show that mental procedures cannot go beyond mechanical procedures. However, this argument is inconclusive. What Turing disregards completely is the fact that mind, in its use, is not static, but constantly developing, i.e., that we understand abstract terms more and more precisely as we go on using them, and that more and more abstract terms enter the sphere of our understanding. There may exist systematic methods of actualizing this development, which could form part of the procedure. Therefore, although

[36]"So I have argued that the Church-Turing Thesis, ... itself seems susceptible to an intuitively compelling demonstration." See Smith [366, p. 366].

[37]Source: In Wikipedia. `en.wikipedia.org/wiki/Alan_Turing`.

[38]Some remarks on the undecidability results [177, pp. 305–307]

at each stage the number and precision of the abstract terms at our disposal may be finite, both (and, therefore, also Turing's number of distinguishable states of mind) may converge toward infinity in the course of the application of the procedure. Note that something like this indeed seems to happen in the process of forming stronger and stronger axioms of infinity in set theory. However, today this process is far from being sufficiently understood to form a well-defined procedure. It must be admitted that the construction of a well-defined procedure which could actually be carried out (and would yield a non-recursive number-theoretic function) would require a substantial advance in our understanding of the basic concepts of mathematics.

Fig. 1.7: E. Post[39]

Not only is the mind not static, but the notion of algorithm keeps evolving, being more open and allowing increased flexibility and expressive power. Some examples include the classical sequential algorithms, parallel, nondeterministic, real-time, analogue, randomised, probabilistic, distributed, quantum, biology-inspired, learning, man-machine algorithms, AI-based algorithms; many others will be invented. Does Church-Turing

[39]Source: In Wikipedia. `en.wikipedia.org/wiki/Emil_Leon_Post#/media/File:Emil_Leon_Post.jpg`.

Thesis cover all computations performed in this multiverse of algorithms? A negative well-argued answer was given by Gurevich [208].

In contrast, Turing and Post thought that Church-Turing Thesis characterises the computational power of humans. Moreover, according to Wolfram [415, p. 1125],

> Turing at least seems to have thought that it might not apply to continuous processes in physics. Kurt Gödel privately discussed the question of whether the universe could be viewed as following Church's Thesis and being 'mechanical'. ... But it was not until the 1980s – perhaps particularly following some of my work – that it began to be more widely realized that Church's Thesis should best be considered a statement about nature and about the kinds of computations that can be done in our universe.

A physical version of the Thesis was proposed independently by Wolfram [414] and Deutsch [159]:

> **Physical Church-Turing Thesis.**[40] Every finite physical system can be simulated to any specified degree of accuracy by a universal Turing machine.

Another variation, known as the Complexity-Theoretic Church-Turing Thesis or the Extended Church-Turing Thesis, refers to what can be "efficiently" computed; this thesis is of interest only if $\mathbf{P} \neq \mathbf{NP}$ [140], and is important for quantum computing [76].

Church-Turing Thesis has been wrongly interpreted in various (sometimes, ingenious) ways [139, Misunderstandings of the Thesis] and here are two examples: i) the universe is equivalent to a Turing machine and incomputable functions are physically impossible and ii) the human mind and/or consciousness are equivalent to and can be instantiated by a Turing machine.[41]

1.13 Is there a quantum Halting Problem?

All computers rely on the ability to store and manipulate information. Classical computers, like Mac or iPad or Apple Watch, store information

[40] Copeland [140] discusses three forms of this Thesis: this form is considered the "bold" one and the other two are "super-bold" and "modest".

[41] Challenge for the reader: Why are i) and ii) wrong?

using classical physical systems, which can be in two states – two different voltage or current levels allowed by a circuit or two distinct levels of light intensity – conventionally named bits and denoted by 0 and 1. Quantum computers use quantum physical systems, which can also be in two states, conventionally named *quantum bits*, or *qubits*, and denoted by $|0\rangle$ and $|1\rangle$; see [103, 299]. A qubit, the quantum analogue of the classical bit, represents the unit of quantum information realised with a two-state quantum device. For example, the polarisation of a single photon in which the two states can be taken to be the vertical polarisation and the horizontal polarisation. In a classical system, a bit is in one state or the other. However, quantum mechanics allows the qubit to be in a superposition of both states simultaneously. A classical analogue is a coin spinning through the air before it lands on a table. Superpositions can be entangled with those of other objects, meaning their outcomes will be "related" even if we don't know yet what they are.

Fig. 1.8: Photograph of a chip constructed by D-Wave Systems Inc., designed to operate as a 128-qubit superconducting adiabatic quantum optimisation processor[42]

The core idea of quantum computation is to hold objects in superposition states long enough to carry out various processes. This is easy to describe mathematically but presents a massive challenge from an engineering point of view because once a superposition meets with materials that are part of a measured system, it loses its in-between state in what is known as *decoherence* and becomes a classical bit. Quantum computers

[42]Source: In Wikipedia. en.wikipedia.org/wiki/D-Wave_Systems#/media/File:DWave_128chip.jpg.

must shield quantum states from decoherence while making them accessible to read and manipulate. Another difficulty is reconciling more robust quantum processes with better ways to check and correct errors.

There are currently two main approaches to physically implementing a quantum computer: analogue and digital. Analogue approaches are divided into quantum simulation, quantum annealing, and adiabatic quantum computing. The most advanced machine in this category is the D-Wave machine, with more than 5,000 qubits and reduced noise using the Pegasus quantum processor chip[43] with 15 connections per qubit [78, 145]. According to [237]:

> Quantum annealing sidesteps the need for extensive error correction, a problem that plagues the gate-level implementations of Google, IBM, Microsoft, Rigetti, and Xanadu, limiting them to fewer than 70 qubits. Only Fujitsu has also adopted quantum annealing, albeit in a 1024-bit digital chip that only emulates a quantum computer.

The QUBO problem consists of the minimisation of a quadratic objective function $f(\mathbf{x}) = \mathbf{x}^T Q\mathbf{x}$, where $\mathbf{x}$ is a n-vector of binary variables and Q is an upper-triangular $n \times n$ matrix of real numbers:

$$x^* = \min_{\mathbf{x}} \sum_{i \leq j} x_i Q_{(i,j)} x_j, \text{ where } x_i \in \{0, 1\}. \tag{1.2}$$

D-Wave solves "natively" only QUBO problems: to solve a problem with this quantum machine, one has to reformulate the problem as a QUBO (as a first step). For example, various methods to construct efficient QUBO formulations for the Graph Isomorphism Problem – one of a very few problems in **NP**[44] that is neither known to be solvable by a polynomial-time algorithm[45] nor **NP**-complete – have been proposed in [89].

Digital quantum computers use quantum (logic) gates to do computation. A quantum gate is a basic quantum circuit operating on a small number of qubits. Quantum gates are the building blocks of quantum circuits, like classical gates are for classical digital circuits.

---

[43]Pegasus used for Advantage QPUs and Zephyr (under development) is the quantum processor for next-generation of QPUs.

[44]See Section 3.9.

[45]An algorithm works in polynomial-time (has polynomial running time) if for some $k, C > 0$, its running time on inputs of size n is at most $C \cdot n^k$.

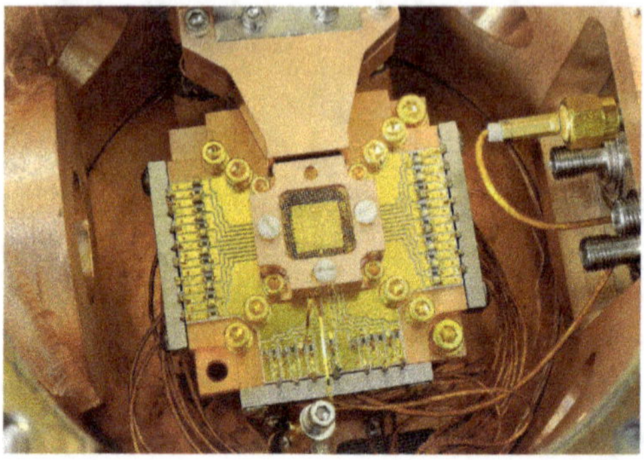

Fig. 1.9: Chip ion trap for quantum computing from 2011 at NIST[46]

Unlike many classical gates, quantum gates are reversible. What does it mean? An operation is *reversible* if it can be *undone*, that is if its inputs can always be deduced from the outputs. Most gates are irreversible, and a typical example is the gate NAND

$$(a, b) \mapsto \neg(a \wedge b) \tag{1.3}$$

which has two input bits and only *one* output bit ($a \wedge b = 1$ if and only if $a = b = 1$ and $\neg(a) = 1 - a$). We cannot recover a unique input from the output bit because 1 can be obtained from three distinct inputs: $(0, 0), (0, 1), (1, 0)$.

To operate a computer, we must ensure that distinct physical states represent distinct logical states. Each bit has two values, so it has one degree of freedom; generally, a set of n bits has n degrees of freedom; they correspond to 2^n physical states. If we erase n bits, say we reset all to 0, then we have compressed 2^n logical states into a single state, a loss of entropy. The irreversible loss of information increases the system's temperature, which determines heat dissipation. Consequently, operations which are not one-to-one, which map distinct logical states into a common one, *cost energy*: they satisfy Landauer's principle [262, 263]:

Erasure of information is a dissipative process.

[46]Source: In Wikipedia. `www.wikiwand.com/en/Trapped_ion_quantum_computer#Media/File:Quantum_Computing;_Ion_Trapping_(5941055642).jpg`, License: Public Domain.

However, performing some classical computing using only reversible gates is possible. A reversible version of the NAND gate is, for example, Toffoli gate[47]

$$(a, b, c) \mapsto (a, b, c \oplus (a \wedge b)). \tag{1.4}$$

Indeed, (1.4) is a reversible 3-bit gate that flips the third bit if the first two both take the value 1 and does nothing otherwise. Hence, the third output bit becomes the NAND of a and b in case $c = 1$. The price paid to get reversibility consisted in adding a new variable c.

input			**output**		
a	b	c	a	b	c
0	0	0	0	0	0
0	1	0	0	1	0
1	0	0	1	0	0
1	1	0	1	1	1

input			**output**		
a	b	c	a	b	c
0	0	1	0	0	1
0	1	1	0	1	1
1	0	1	1	0	1
1	1	1	1	1	0

Table 1.1: Toffoli's gate

Similar ideas can be used to produce reversible versions of the binary addition, $(a, b) \mapsto (a, a \oplus b, a \wedge b)$ and real addition $(x, y) \mapsto (x+y, x-y)$. In the first case, we replicated the first variable a; in the second case, we added a new component storing additional value. Note that a reversible gate has to have the same number of input and output variables (to guarantee bijectivity).

For one input bit, there are two possible reversible gates. One of them is NOT. The other is the identity gate, which maps its input to the output unchanged. The only non-trivial reversible gate for two input bits is the NAND gate. The Toffoli gate is *universal* in the sense that for any Boolean function $f(x_1, x_2, \ldots, x_m)$, there is a circuit consisting of Toffoli gates that takes $(x_1, x_2, \ldots, x_m)$ and some extra bits set to 0 or 1 to outputs $(x_1, x_2, \ldots, x_m), f(x_1, x_2, \ldots, x_m)$ plus some extra bits (called garbage). This means that every Boolean function can be computed reversibly by a system built from Toffoli gates [396].

On 19 October 1900, Max Planck, a 42-year-old physicist, presented to a meeting of the German Physical Society, a ground-breaking solution to a

[47]Recall that $a \oplus b$ is 1 if and only if $a \neq b$.

long-standing problem: Why does the colour of radiation from any glowing body change from red to orange and ultimately to blue as its temperature increases? Planck based his answer on the assumption that radiation (like matter) comes in *discrete quantities* of energy, the so-called "quanta".[48] The quantisation of energy proved to be a revolutionary and fundamental hypothesis, with extraordinary conclusions contradicting what common sense imposed: things change because you "look" at them, an action can precede its cause (Compton effect), objects have "random" behaviour, etc.[49]

All knowledge in the quantum world is based upon observing qubits. The information derived from an elementary act of observation of a qubit is no more than a single bit, but *there is more on it than that.* To model a qubit mathematically, we need to use a 2-dimensional space and adopt a standard system of reference (an orthonormal basis of vectors):

$$|0\rangle = \begin{pmatrix} 1 \\ 0 \end{pmatrix}, \ |1\rangle = \begin{pmatrix} 0 \\ 1 \end{pmatrix},$$

so that every qubit $|x\rangle$ can be represented uniquely as a superposition of $\{|0\rangle$ and $|1\rangle\}$:

$$|x\rangle = a|0\rangle + b|1\rangle = \begin{pmatrix} a \\ b \end{pmatrix}, \tag{1.5}$$

where a, b are either real (or complex numbers) with $|a|^2 + |b|^2 = 1$ (for normalisation reason). Intuitively, $|x\rangle$ has some $|0\rangle$ and $|1\rangle$ proportionally with a and b, respectively.

We can perform a measurement that projects the qubit onto the basis $\{|0\rangle, |1\rangle\}$ and we obtain the outcome $|1\rangle$ with probability $|b|^2$, and the outcome $|0\rangle$ with probability $|a|^2$. For example, measuring the qubit

$$|x\rangle = \frac{1}{\sqrt{2}}(|0\rangle + |1\rangle)$$

or the qubit

$$|y\rangle = \frac{1}{\sqrt{2}}(|0\rangle - |1\rangle)$$

[48] The Latin for amount.

[49] Planck felt very uncomfortable with the idea that quantisation is a law of nature. In this connection, he is notorious for having said that "new scientific theories supplant previous ones not because people change their minds but simply because old people die." But are there any laws of nature? According to an unconventional view [100, 109], the notion of "law of nature" or "the law of the universe" is at least misleading if not vacuous.

we get $|0\rangle$ and $|1\rangle$ with equal probability, i.e. $\frac{1}{2}$. This is the simplest form of obtaining quantum randomness, a topic we will discuss more in Section 4.12.

Except limit cases $a = 0$ and $b = 0$, the *measurement irrevocably disturbs the state*: If the value of the qubit is initially unknown, then there is no way to determine a and b with any conceivable measurement. However, *after* performing the measurement, the qubit is in a known state (either $|0\rangle$ or $|1\rangle$; these special qubits correspond to the classical bits 0 and 1, respectively).

The ability of quantum systems to exist in a "blend" of all their allowed states simultaneously is guaranteed by the *Principle of Superposition.* Even though a qubit is in a superposition (1.5), it contains no more information than a classical bit, despite its having "temporarily" infinitely many states. The reason is that *information can be extracted only by measurement.* But, as we have shown, for any measurement of a qubit with respect to a given orthonormal basis, $\{|0\rangle, |1\rangle\}$, there are only two possible results, corresponding to the two vectors of the basis. On the other hand, capturing more information by measuring on two different bases is impossible because the measurement changes the state. Even worse, quantum states cannot be cloned; hence it's impossible to measure a qubit in two different ways (even indirectly, by using a copy trick, that is, copying and measuring the copy) – this is the Non-Cloning Theorem. However, a universal cloning machine can make a clone of an unknown state with a surprisingly high fidelity of 5/6; see [55].

The quantum evolution of a qubit $|x\rangle$ is described by a 2×2 matrix, which can be considered a quantum gate. The simplest example is the quantum gate NOT, which interchanges the vectors $|0\rangle$ and $|1\rangle$ and is given by the matrix

$$\text{NOT} = \begin{pmatrix} 0 & 1 \\ 1 & 0 \end{pmatrix}.$$

It flips that state of its input,

$$\text{NOT}\ |0\rangle = \begin{pmatrix} 0 & 1 \\ 1 & 0 \end{pmatrix} \begin{pmatrix} 1 \\ 0 \end{pmatrix} = \begin{pmatrix} 0 \\ 1 \end{pmatrix} = |1\rangle,$$

and

$$\text{NOT}\ |1\rangle = \begin{pmatrix} 0 & 1 \\ 1 & 0 \end{pmatrix} \begin{pmatrix} 0 \\ 1 \end{pmatrix} = \begin{pmatrix} 1 \\ 0 \end{pmatrix} = |0\rangle$$

Classical gates have corresponding quantum ones, but some quantum gates have no classical counterpart. The square-root of NOT (introduced by Deutsch [160]) given by the matrix ($i = \sqrt{-1}$):

$$\sqrt{\mathrm{NOT}} = \frac{1}{2}\begin{pmatrix} 1+i & 1-i \\ 1-i & 1+i \end{pmatrix}$$

satisfies the equality

$$\sqrt{\mathrm{NOT}} \cdot \sqrt{\mathrm{NOT}} = \mathrm{NOT},$$

but *not single-input/single-output classical binary gate can satisfy it.*[50]

We can now ask the question: what is the computational power of digital quantum computers? More precisely, can a digital quantum computer simulate exactly a universal Turing machine? *The answer is negative.* The argument is simple: every universal Turing machine computes a partially defined function, that a function which is undefined for some arguments (actually, for infinitely many arguments), but quantum algorithms built with quantum gates cannot compute such a function because they compute only functions defined in every argument [18]. A simple example of a partial function no quantum algorithm can compute is the partial function of two natural arguments (n, m), which outputs the integer part of the fraction $\frac{n}{m}$ if $m > 0$ and is undefined for all inputs $(n, 0)$. Here are two significant consequences:

(1) quantum computing potential advantages could come only from faster than classical computations,
(2) all quantum algorithms for digital and analogue quantum computers halt on every input, so the Halting Problem is trivial in quantum computing.

In contrast with the second consequence, the first one requires more discussion. How fast could quantum computers solve (some classes of) problems? A largely-spread view is [235]:

> A hyper-fast quantum computer is the digital equivalent of a nuclear bomb; whoever possesses one will be able to shred any encryption and break any code in existence.

The quotation above comes from a mystery novel, but the myth of *quantum supremacy* is largely accepted despite being at best highly problematic [76]. The media has helped to convince politicians, funding agencies and the large public that the quantum revolution "is at our doorstep" and with it, the major problems of humanity will be quickly and definitively solved. The following citation from an article authored by an IBM brand

[50]Challenge for the reader: Prove it.

contributor, a company itself involved in the race for quantum supremacy, offers a glimpse of "soon coming" quantum miracles [231]:

> Hyper-accurate long-term weather forecasting. Life-saving drugs discovered through deep study of the behavior of complex molecules. New synthetic carbon-capturing materials to help reverse climate change caused by fossil fuels. Stable, long-lasting batteries to power electric vehicles and store green energy for the utility grid.

Before discussing quantum miracles, let us pause and ask a simple question: when will useful quantum computers be constructed? The answer repeated in the last two decades by the most optimistic experts was about ten years. More cautious ones predict 20 to 30 years, i.e. around 2045 to 2065. In contrast, physicist M. I. Dyakonov argued in his recent book [168]:

> Not in the foreseeable future.

Is this a singular opinion? The answer is negative. Over time a minority of "quantum computing sceptics" have discussed mathematical, physical and engineering arguments against the feasibility of constructing quickly useful quantum computers. A short selection includes papers by Nobel Prize winners (for physics) S. Haroche [213], Gerard't Hooft [228] and R. Laughlin [265], multi-awarded mathematician G. Kalai [241, 242], IBM physicist R. Landauer author of the principle bearing his name [264], physicist and computer scientist S. Wolfram author of *Mathematica* [414].

It is appropriate to close this section with the disclaimer R. Landauer suggested decades ago. Add to all papers in quantum computing the following sentence:

> This scheme, like all other schemes for quantum computation, relies on speculative technology, does not in its current form, take into account all possible sources of noise, unreliability and manufacturing error, and probably will not work.

1.14 Programming languages with decidable Halting Problem

Pseudocode programs have a powerful property: Turing completeness or universality. A programing language is called *Turing complete* or *computationally universal* if it can simulate any Turing machine.[51] In detail, for every Turing machine M, there exists a program P such that for every input x, $P(x) = M(x)$. This means that if $M(x)$ halts, then $P(x)$ halts and $P(x) = M(x)$; otherwise, $P(x)$ does not halt. As always, we assume that the program M has enough time and memory to run on input x.

The Halting Theorem in Section 1.10 indirectly uses Turing's completeness;[52] hence every programming language that can simulate all pseudocode programs is subject to the Halting Theorem. Virtually all programming languages today (e.g. Java, JavaScript, Perl, etc.) are Turing complete because they each implement all the features required to run pseudocode programs.

However, not all programming languages are Turing complete. In the late 19th century, Kronecker formulated the first notions of computability, including primitive recursion [258] and a class of functions based on it. Primitive recursive functions can be computed by programs with bounded recursion only, i.e. by programs using only loops for which an upper bound of the number of iterations can be determined before entering the loop.[53] Every programming language computing exactly the class of primitive recursive functions is not Turing complete because its instructions do not allow infinite computations, particularly infinite loops. This fact also shows that all primitive functions are total; they are defined on every input. In particular, the Halting Problem is trivial for this class, like for the class of quantum programs 1.13.[54] Early versions of the GLSL and HLSL programming languages (used in GPUs for rendering graphics) prohibited unbounded loops and enforced those rules in their respective compilers, which made it impossible to write a program that didn't halt.

1.15 Enumerators

An *enumerator* is a program that lists, possibly with repetitions, elements of some set, which it is said to enumerate. A set enumerated by some

[51] See Sections 1.1 and 5.2 for more details.
[52] Challenge for the reader: Why?
[53] These loops are sometimes called "for loops". See [60].
[54] Can quantum computing compute total non-primitive recursive functions?

enumerator is called *computably enumerable*[55].

For example, the program prints the sequence $1, 2, \ldots, N$:

```
PROGRAM(Enum, N)
read N
i = 1
while i ≤ N do
    print i
    i = i + 1
end while
stop
```

The program

```
PROGRAM(Enum all)
i = 1
while i ≥ 1 do
    print i
    i = i + 1
end while
stop
```

prints the sequence of all positive integers: $1, 2, \ldots$ and the program

```
PROGRAM(Enum all − repeat)
i = 1
while i ≥ 1 do
    print i
    print i
    i = i + 1
end while
stop
```

prints the sequence of all positive integers with repetitions: $1, 1, 2, 2, \ldots$

The class of computably enumerable sets is closed under union and intersection but not under complement. Every decidable set is computably enumerable, but the converse implication is false; examples of computably enumerable but undecidable sets will be given in Section 2.2. Finally,

[55] Sometimes also called *recursively enumerable*, *(Turing) recognisable*.

Decidability Theorem. *A set is decidable if and only if the set and its complement are both computably enumerable.*

A proof of this result will be given in Section 5.4.

1.16 Approximate solutions to the Halting Problem

Can the Halting Problem be approximately solved? Answers have been proposed by various authors [285, 107, 288, 85, 44, 91, 92]. In this section, we present a solution expressed as an anytime algorithm. Following [205], an anytime algorithm trades execution time for quality of results; the algorithm returns not only a result but also a "quality measure" which evaluates how close the obtained result is to the result that would be returned if the algorithm ran until completion (which may be prohibitively long or even infinite, as proposed by Manin in [288]). To improve the quality of the solution, the anytime algorithm can be continued after it has halted, if the output is not considered acceptable.

Fig. 1.10: Y. I. Manin[56]

The idea of the anytime algorithm solving the Halting Problem is the following. We fix a rational number ε in $(0, 1)$ in terms of which we define the "quality measure" of the decision. Then we construct a program which stops on every program p (as input) and outputs one of the following decisions: a) p *halts*, and in this case, the result is provably correct, or b) p

[56]Source: In Wikipedia. `tinyurl.com/5et97rry`.

does not halt, and in this case, the probability (adequately defined) that the result is wrong is less than ε.

Running times play an important role in this problem because halting programs are not uniformly distributed: see [425, 423, 370] for experimental work and [252, 85] for theoretical results, in particular the result discussed in Section 5.11 that every program either stops "quickly" or never stops [107].

The statistical anytime algorithm has two parts: the pre-processing which is done only once (when the parameters of the quality of solutions are fixed) and the central part, which is run for any input program. With a confidence level as large as required, the algorithm produces correct decisions with a probability as large as requested.

Using an appropriate statistical framework, we construct an anytime statistical algorithm which uses three parameters: the probability of an erroneous decision, the precision of and the confidence level in the estimation. The input program – tested for termination – is run for up to the largest number of steps made by any of the sampled programs. If the computation does not terminate by this time threshold (cut-off), then the (possibly wrong) decision is that the program will never stop. With a confidence level as large as required, the anytime statistical algorithm produces correct decisions with a probability as large as required.

The assumption is that the probability distribution of running times of programs is unknown.

The algorithm operates with three parameters:

a) a *bound* $\varepsilon \in (0,1)$ for the decision error,
b) a *precision* parameter $1 < \lambda < \varepsilon$, and
c) a *confidence* parameter $1-\delta \in (0,1)$ which is a probabilistic bound on the confidence in the precision parameter.

We sample N independent halting programs $x_1, \ldots, x_N \in \mathrm{dom}(\mathbf{U})$, where $\mathbf{U}$ is the universal Turing machine[57] and by running them on $\mathbf{U}$ till they stop, we calculate their respective running times $t(\mathbf{x}) = (t_1(\mathbf{x}), \ldots, t_N(\mathbf{x}))$. Randomisation is done according to the probability distribution induced by an injective computable enumeration of the halting programs.

The statistical anytime algorithm is:[58]

---

[57]Remember that $\mathrm{dom}(\mathbf{U})$ is not computable.
[58]The integer part of the real r is denoted by $\lceil r \rceil$.

Pre-processing.
Fix three rational numbers $\varepsilon, \lambda, \delta \in (0,1)$ *with* $\lambda < \varepsilon$.
Compute $N = N(\lambda, \delta) = \lceil \frac{1}{2\lambda^2} \cdot \ln \frac{1}{\delta} \rceil$.
Sample N *independent programs* $\mathbf{x} = (x_1, \ldots, x_N) \in \mathrm{dom}(\mathbf{U})^N$ *and calculate their respective running times* $\mathbf{t}(\mathbf{x}) = (t_1(\mathbf{x}), \ldots, t_N(\mathbf{x}))$.
Increasingly order the observed running times $t_i(\mathbf{x})$ *and compute the threshold* $\mathbf{T} = t_{(\lceil N(1-\varepsilon+\lambda) \rceil)}(\mathbf{x})$.
Main part.
Let z *be an arbitrary program for* $\mathbf{U}$.
If the computation $\mathbf{U}(z)$ *does not stop in time less than or equal to* $\mathbf{T}$, *then declare that* $\mathbf{U}(z) = \infty$.

In [92], it is proven that the probability of the event that the statistical anytime algorithm makes a wrong decision, that is, it declares $\mathbf{U}(z) = \infty$ when there exists $t > t_{(\lceil N(1-\varepsilon+\lambda) \rceil)}$ such that $\mathbf{U}(z)$ stops in time t, is smaller or equal to ε, is larger than $1 - \delta$, i.e. the probability of error is smaller than or equal to ε with confidence larger than $1 - \delta$.

The standard implementation of the statistical anytime algorithm is as follows. Given three rational numbers $\varepsilon, \lambda, \delta \in (0,1)$ with $\lambda < \varepsilon$, first compute the sample size $N = \lceil \frac{1}{2\lambda^2} \cdot \ln \frac{1}{\delta} \rceil$; this positive integer is fixed as long as λ, δ are fixed. Then use an algorithm to generate a random injective computable enumeration of $\mathrm{dom}(\mathbf{U})$ till N programs $x_1, \ldots, x_N$ and their running times $t_1, \ldots, t_N$ are obtained; again, these programs are fixed with $\varepsilon, \lambda, \delta$. Then, for every program $z \in \mathbf{N} \setminus \{0\}$, if the computation $\mathbf{U}(z)$ does not stop in time $t_{(\lceil N(1-\varepsilon+\lambda) \rceil)}$, then declare that $\mathbf{U}(z) = \infty$. In the latter case, the probability of error is smaller than or equal to ε with confidence larger than $1 - \delta$.

In a second implementation, we start with two rational numbers $\varepsilon, \lambda \in (0,1)$ with $\lambda < \varepsilon$ and an "affordable" size $\tilde{N}$ of samples (programs and running times), then compute the rational $\delta(\tilde{N}, \lambda) \in (0,1)$ satisfying the inequality $N \geq N(\lambda, \delta) = \lceil \frac{1}{2\lambda^2} \cdot \ln \frac{1}{\delta} \rceil$; see [92]. We continue with the standard implementation of the statistical anytime algorithm with parameters ε, λ $(\lambda < \varepsilon)$ and

$$\delta(\tilde{N}, \lambda) = \exp\left(-2\tilde{N} \cdot \lambda^2\right) \tag{1.6}$$

to calculate the size sample $N(\lambda, \delta(\tilde{N}, \lambda)) = \tilde{N}$. However, the value $\delta(\tilde{N}, \lambda)$

in (1.6) is not rational, so we need to calculate a rational approximation

$$\delta(\tilde{N}, \lambda) \geq \exp\left(-2\tilde{N} \cdot \lambda^2\right),$$

which implies the inequality $N(\lambda, \delta(\tilde{N}, \lambda)) < \tilde{N} + 1$.

δ	ε	$\lambda\,(< \varepsilon)$	$N\,(\lambda, \delta)$	$\lceil N\,(\lambda, \delta) \cdot (1 - \varepsilon + \lambda) \rceil$
$\frac{1}{100}$	$\frac{5}{1000}$	$\frac{1}{1000}$	2.3026×10^6	2.2934×10^6
$\frac{1}{100}$	$\frac{1}{1000}$	$\frac{5}{10000}$	9.2103×10^6	9.2057×10^6
$\frac{5}{1000}$	$\frac{5}{1000}$	$\frac{1}{1000}$	2.6492×10^6	2.6386×10^6
$\frac{5}{1000}$	$\frac{1}{1000}$	$\frac{5}{10000}$	1.0597×10^7	1.0592×10^7
$\frac{1}{1000}$	$\frac{5}{1000}$	$\frac{1}{1000}$	3.4539×10^6	3.4401×10^6
$\frac{1}{1000}$	$\frac{1}{1000}$	$\frac{5}{10000}$	1.3816×10^7	1.3809×10^7

Table 1.2: Numerical illustration of the first implementation of the anytime algorithm

The "price" paid working with an affordable, smaller sample size $\tilde{N}$ is a (possibly sharp) decrease in the confidence level.

In a third implementation, we start with two rational numbers $\varepsilon, \lambda \in (0, 1)$ with $\lambda < \varepsilon$ and an "affordable" upper bound T on the running time of the computation $\mathbf{U}(z)$ in the statistical anytime algorithm. We then use an injective dovetailing algorithm to generate as many elements of dom($\mathbf{U}$) as possible in time T. In this way, we will obtain a sample of $N(T)$ programs $x_1, \ldots, x_{N(T)}$ and their respective running times $t_1, \ldots, t_{N(T)}$ such that each program stops in time at least (in fact, much smaller than) T: $t_i \leq T$, for all $1 \leq i \leq N(T)$. We then continue with the second implementation with parameters $\varepsilon, \lambda, \tilde{N} = N(T)$.

ε	$\lambda(<\varepsilon)$	$\tilde{N}$	$\delta\left(\tilde{N}, \lambda\right)$	$\left\lceil \tilde{N}(1-\varepsilon+\lambda) \right\rceil$
$\frac{1}{100}$	$\frac{5}{1000}$	10^5	6.7379×10^{-3} (very good)	9.95×10^4
$\frac{1}{100}$	$\frac{4}{1000}$	10^5	4.0762×10^{-2} (good)	9.94×10^4
$\frac{1}{100}$	$\frac{5}{1000}$	$2 \cdot 10^5$	4.54×10^{-5} (excellent)	1.99×10^5
$\frac{1}{100}$	$\frac{4}{1000}$	$2 \cdot 10^5$	1.6616×10^{-3} (very good)	1.988×10^5
$\frac{1}{100}$	$\frac{1}{1000}$	10^6	1.3534×10^{-1} (hardly acceptable)	9.91×10^5
$\frac{1}{1000}$	$\frac{5}{10000}$	10^6	6.0653×10^{-1} (unacceptable)	9.995×10^5

Table 1.3: Numerical illustration of the second implementation of the anytime algorithm

The second and third implementations can be improved by increasing the sample size or the time-bound, respectively. The above two tables illustrate numerically the first and second implementation methods.

Approximations in algorithmic information theory [61], for example, the approximations of Solomonoff universal distribution, a *apriori* incomputable probability distribution over the set of finite binary strings [274], involve constants; in contrast, all implementations of the statistical anytime algorithm are free from this uncertainty.

The article [267] combines an exhaustive exploration of programs with an estimation of their non-halting probability to produce approximations of algorithmic complexity.

Chapter 2

Incompleteness and Halting

2.1 Euclid, Hilbert and the axiomatic method

For over two thousand years, Euclid's *Elements* had provided the model of mathematical rigour and the link between the geometric constructions in mathematical spaces – modelled from the world of senses and actions – and the physical space, at any scale. The theorems proven in the mathematical space model well phenomena in the physical space, like the relationships between stars and between Democritus's atoms. Euclid's model transgresses mathematics: for example, B. Spinoza presented moral and religious arguments in the Euclidean style in his book *Ethics, Demonstrated in Geometrical Order* published posthumously in 1677.

Euclid was probably the first mathematician to use the *axiomatic method* (in his geometry). In the modern era, this method consists of a formal language constructed according to well-defined rules, together with some basic notions and statements – called *axioms or postulates* – and *deduction rules*. An axiom or a postulate is a statement that asserts a certain property of a basic notion, or that a certain relation holds between some basic notions. The axioms are accepted.[1] From them, new notions are defined, and new statements are *deduced* or *proven* using the deduction/inference rules.[2] A *proof* is a finite sequence of statements each of which is either an axiom or can be deduced by some inference rule from

[1]If the axioms are in "perfect" agreement with human experience in the particular field of interest – as it was required from Euclid to the mid-nineteenth century – their validity is beyond any doubt. Thus, no proofs of axioms are required; they are *evident*.

[2]A common deduction rule is *modus ponens*, Latin for "affirming affirms", "method of putting by placing", first used by Theophrastus (c. 371 – c. 287 BC): "P implies Q and P is asserted to be true, therefore Q must be true". See `en.wikipedia.org/wiki/Modus_ponens`.

previous statements in the sequence. A statement for which there exists a proof is called a *theorem*. An important requirement is that *the set of axioms is either finite or can be listed by a program.* The definition of proof and the above assumption allow that each step of the purported proof can be algorithmically checked for correctness by a *proof-checking program* (as Hilbert required), i.e. the set of proofs is decidable. Furthermore, they also imply that *the set of theorems can also be listed by a program.* The propositional logic, the predicate logic and Peano (axiomatic) arithmetic are examples of axiomatic (or formal) systems with decidable sets of proofs; they are discussed in Chapter 6.

Fig. 2.1: Euclid Statue in Oxford[3]

The goal in choosing an axiomatic (formal) system – a problem in mathematical modelling – is to be able to (rigorously) *prove as many correct results as possible in a particular field of interest, without proving any incorrect results.*[4] For example, an axiomatic system for arithmetic should prove as many, ideally all, true statements about natural numbers without proving any incorrect results. But how do we know that a statement about natural numbers is true? Well, such a statement is true if it has been

[3]Source: In Wikipedia. `simple.wikipedia.org/wiki/Euclid`.

[4]Formalisation is difficult, and many proposals of axiomatic systems were initially contradictory [279, p. 10].

proven... Is this not a circularity? This is a delicate issue. Most true statements in arithmetic are proven in an ad-hoc manner, not in an axiomatic way. To have a solid foundation for a more rigorous notion of proof, an adequate framework was required: this was provided by the new *mathematical logic* developed since the mid-nineteenth century by the efforts of English mathematician G. Boole, German philosopher, logician, and mathematician G. Frege, Italian mathematician G. Peano, German mathematicians R. Dedekind and D. Hilbert, English mathematician and logician B. Russell, Austrian logician, mathematician and analytic philosopher K. Gödel and English mathematician and computer scientist A. Turing, and a few others [330].

Fig. 2.2: D. Hilbert[5]

D. Hilbert, one of the greatest mathematicians of the nineteenth century, believed that an axiomatic theory could be developed independently of any need for intuition and would bring clarity and rigour to the extent that the verification of proof could be done by a mechanical device.[6] Hilbert wrote a (hugely influential) textbook *Foundations of Geometry* (1899) to promote his vision of an axiomatic (Euclidean) geometry. One of the pillars of this

[5]Source: In Wikipedia. en.wikipedia.org/wiki/David_Hilbert.

[6]In the practice of mathematics this level of detail is rarely attained. Later on, we will discuss more formal proofs.

geometry is the parallel postulate, also called *Euclid's fifth postulate*, which in the two-dimensional geometry reads:

> If a line segment intersects two straight lines forming two interior angles on the same side that sum to less than two right angles, then the two lines, if extended indefinitely, meet on that side on which the sum of the angles to less than two right angles.

This postulate is equivalent to the more known statement that the sum of internal angles of any triangle is exactly two right angles, that is, 180°.

Fig. 2.3: B. Riemann[7]

In the nineteenth century, doubts arose about whether axioms are always "evident". A specific challenge affected Euclid's axiomatic treatment of geometry. For Galileo, Kepler, Newton and Kant, the physical universe was faithfully described by the language of Euclidean circles, triangles and straight lines. But this "belief" was shaken by the German mathematician B. Riemann, who in his 1854 *Habilitation Thesis* proposed a new geometry – later called non-Euclidean – in which the sum of the internal angles of a triangle of stellar orders of magnitude can be more than 180°, in stark contradiction with Euclid's fifth postulate. Moreover, Riemann

[7]Source: In Wikipedia. `en.wikipedia.org/wiki/Bernhard_Riemann`.

boldly conjectured that the "forces of cohesion between bodies are linked to the metric of space" and used it to link the metric to the curvature[8] of the space. Later, Einstein gave a specific physical interpretation to these mathematical results. Locally, Euclidean geometry works well for planes of null curvature; however, on the scale of the universe, it is precisely the non-null curvature which enables the unification between gravitation and inertia, the crux of Einstein's relativity. This "delirium" forced Riemann and his followers to abandon the Cartesian space, so intuitive in daily life, in favour of non-Euclidean geometries that can be obtained by negating the parallel postulate. This was a sensational shock: *the axiomatic method could not be unconditionally trusted anymore because long-term experience and intuition could be misleading.* To be trusted, the axiomatic method itself had to be studied and based on solid foundations [367, 279].

Fig. 2.4: A. Einstein[9]

The *consistency* of the axiomatic system, that is, the impossibility of proving (in the system) a statement and its negation, is of utmost impor-

[8]The curvature is the amount by which a curve deviates from being a straight line, or a surface deviates from being a plane.

[9]Source: In Wikipedia. `upload.wikimedia.org/wikipedia/commons/7/78/Einstein_1921_by_F_Schmutzer_-_restoration.png`.

tance for the axiomatic treatment.[10] In Hilbert's words (from the lecture "Axiomatic Thought" presented to the Swiss Mathematical Society (1917)):

> The chief requirement of the theory of axioms must go farther [than merely avoiding known paradoxes][11], namely, to show that within every field of knowledge contradictions based on the underlying axiom-system are absolutely impossible.

Hilbert's program proposed in 1920–1921 the following "solution": *ground all existing mathematical theories on a finite, complete and consistent axiomatic system.* Informally, this calls for

- a formalisation of mathematics, that is, a precise formal language based on well-defined rules, in which all mathematical statements are written,
- a finite set of axioms (axiom schemes) and rules of inference,
- a proof that all *true* mathematical statements can be proven in the adopted axiomatic system *(semantic completeness)*,
- a proof – preferably using only finite mathematical objects – that no contradiction can be obtained in the axiomatic system *(consistency)*,
- an algorithm for deciding the truth or falsity of every mathematical statement (*decidability*).

The consistency of more complicated systems could be proven in terms of simpler systems, and, ultimately, the consistency of all mathematics will be reduced to that of arithmetic.

The new mathematical logic and the impressive three volumes of Russell and Whitehead *Principia Mathematica* [350] provided the logical basis for the development of Hilbert's program, which by 1920 enlisted the works of important mathematicians and logicians such as P. Bernays, W. Ackermann, J. von Neumann and J. Herbrand.

At the International Congress of Mathematics held in Paris in 1900, Hilbert gave a lecture modestly titled "Mathematical Problems" in which he discussed 23 open problems, among them the problem of consistency of arithmetic. In the talk – which proved to be of lasting importance for the

[10] In most, but not all, axiomatic systems, *contradictions entail everything,* that is, all formulae of the system are provable [272].

[11] Like those discovered in G. Cantor's naive set theory by B. Russell and many others which provoked a mathematical foundational crisis (in German, Grundlagenkrise der Mathematik). Russell's paradox showed the inconsistency of "the set of all sets that are not members of themselves".

mathematics of the new century – Hilbert expressed the deep conviction in the solvability of all mathematical problems, phrased as an *axiom*:

> Is the axiom of solvability of every problem a peculiar characteristic of mathematical thought alone, or is it possibly a general law inherent in the nature of the mind, that all questions which it asks must be answerable? ... This conviction of the solvability of every mathematical problem is a powerful incentive to the worker. We hear within us the perpetual call: There is the problem. Seek its solution. You can find it by pure reason, for in mathematics there is no *ignorabimus*.

Fig. 2.5: B. Russell[12]

Was Hilbert's vision just a mathematical dream? The step-by-step method to prove consistency echoes the mechanistic reduction proposed by the French mathematician and physicist P.-S. Laplace more than a century before and adopted by the physicists of the eighteenth and nineteenth centuries. According to this vision, the hypothesis of unity and stability

[12]Source: In Wikipedia. `upload.wikimedia.org/wikipedia/commons/5/5f/Bertrand_Russell_1957.jpg`.

of the universe imposes an agreement of method, which means the identity between macroscopic and microscopic physical laws, all reducible to the elementary ontology underlying matter, movement and force. In this way *the laws of physics, expressed in various equations, will describe any physical event allowing accurate predictions.* Theory and practice will be intimately linked: theories will be developed, predictions will be made and then compared against observations and measurements to validate the theories. *Exceptions* have been found in the form of some unpredictable evolutions (like the throws of dice). Still, they have been considered rare. Even more importantly, it was hoped that their predictability could be restored in terms of probability theory (to which Laplace himself made significant contributions). However, Laplace's Hypothesis that

> [i]t must be possible to deduce all astronomical facts ...

Fig. 2.6: H. Poincaré[13]

didn't last long as in 1892, the French mathematician H. Poincaré showed that in the Three-Body Problem (say, the Sun, Moon and Earth) the weak gravitational interaction effects can add up in unpredictable ways, so:

[13]Source: In Wikipedia. `upload.wikimedia.org/wikipedia/commons/f/f4/PSM_V82_D416_Henri_Poincare.png`.

prediction becomes impossible ...

This discovery sharply contradicted the belief in the solar system's stability, confirmed empirically by historical observations. How to reconcile these facts?

The great I. Newton proposed a solution to guarantee the system's stability "in saecula saeculorum" (Latin for "in an age of ages"): *once in a while, skilful adjustments by God reestablish order.* But this theological solution was rejected by Laplace, who believed "that a thorough mathematical analysis should demonstrate the stability of the system and its full predictability" [279, p. 6].

Fig. 2.7: I. Newton[14]

However, there is no solution as *the solar system is unpredictable* [51]. It may come as a surprise, but deterministic systems can be unpredictable, *chaotic* in modern terms. Throwing a die is the simplest such example: the system can be described by deterministic equations, but it is unpredictable because of the high sensitivity to initial conditions [161]. Poincaré was laying the foundations of chaos theory [51, 365] by showing that the solar system is chaotic.

[14]Source: In Wikipedia. `upload.wikimedia.org/wikipedia/commons/3/39/GodfreyKneller-IsaacNewton-1689.jpg`.

A famous example of a butterfly's wings in Brazil causing a tornado in Texas is a well-known example[15] illustrating how small initial changes in a nonlinear system can lead to significant differences in future states [365]. The mathematician B. Mandelbrot, the "father of fractals", applied the butterfly effect to economics [287].

Hilbert's program motivated the exceptional work of the young mathematician and logician K. Gödel. Informally, his *Incompleteness Theorem* says that

> In every axiomatic system satisfying certain conditions, there exist statements that themselves and their negations are unprovable.

The conditions require [366] that the axiomatic system is

(CEN) *computably enumerable*, i.e. the set of axioms (hence theorems) can be enumerated by a program,
(CON) *consistent*, i.e. the system cannot prove both a statement and its negation, and
(CAR) *rich enough*, i.e. the system contains a modicum of arithmetic of natural numbers.

If (CEN), (CON) and (CAR) are satisfied, then there exists (and can be effectively constructed) a statement G such that both G and its negation non-G are unprovable in the axiomatic system, so the condition:

(SCO) the axiomatic system is *syntactically complete*,[16]

is false: *the axiomatic system is syntactically incomplete.* Gödel's Incompleteness Theorem and its proof [194] are purely syntactic, in the sense that they do not use the semantic notion of "truth". If the axiomatic system has "sound semantics", then the Incompleteness Theorem can be re-phrased:

> In every computably enumerable, rich enough and consistent axiomatic system, there exists a true but unprovable sentence; that is, the axiomatic system is semantically incomplete.

[15] The butterfly effect was discovered by mathematician and meteorologist E. N. Lorenz in 1961.
[16] That is, for any statement, either the statement or its negation is provable in the system.

2.2 From the Halting Theorem to the Semantic Incompleteness Theorem

The Halting and Incompleteness Theorems express the impossibility of dealing algorithmically with some forms of infinity. A deeper relationship can be established. To illustrate the difference between the Halting Theorem and a form of Semantic Incompleteness Theorem, we prove, following [149], a form of the Incompleteness Theorem using the Halting Theorem.

By $N(P, v)$ we denote the statement

the program[17] P will never halt on input v.

For any particular program P and input v, $N(P, v)$ is a perfectly definite statement which is either true (if P does not halt) or false (if P eventually halts). The falsity of $N(P, v)$ can be demonstrated by running P on v. As we know, no amount of computation will suffice to show that every true statement $N(P, v)$ is true: we may still be able to prove by a logical analysis of some P's behaviour that $N(P, v)$ is true, but, because of the Halting Theorem, no such method can work correctly in *all* cases.

Can we choose an *axiomatic system in which all true statements* $N(P, v)$ *and only them are provable?* We assume that the axiomatic system can formalise the statements $N(P, v)$, where P is a program and v is an input for P, the set of proofs is decidable (hence it satisfies (CEN)) and is:

Sound: For all P, v, if there is a proof Π for $N(P, v)$, then P never halts on input v.

Semantically complete: For all P, v, if P never halts on input v, then there is a proof Π for $N(P, v)$.

Soundness means that the system can only prove "true" statements $N(P, v)$. Semantical completeness states the converse: every true statement $N(P, v)$ is provable.

We will prove that an affirmative answer will contradict the Halting Theorem (see more in Section 6.3), so the answer is *negative*:

Incompleteness Theorem. (Incompleteness Semantic Variant 1) *No axiomatic system in which all statements of the form* $N(P, v)$ *are formalised and the set of proofs is decidable, can be both sound and semantically complete.*

[17] Recall that program means pseudocode program, a programming language as powerful as Turing machines, Section 1.3.

Here is a proof. Because the set of proofs is decidable, we can construct a *proof-checking program* that can determine whether an alleged proof Π for $N(P, v)$ is a correct proof in the system.

Suppose, by absurdity, there exists such a *sound and syntactically complete axiomatic system.* Suppose that the proofs in this system are particular strings of symbols on the specific finite alphabet of the system, and let $\Pi_1, \Pi_2, \Pi_3, \ldots$ be the shortlex enumeration of all finite strings on this alphabet.[18] This enumeration is computable and includes all possible proofs in the system and many other strings, including many meaningless strings.

Next, we show how to use this sound and the semantically complete axiomatic system to solve the Halting Problem.

To achieve this goal, we use the following algorithm to test whether a program P halts on the input v. The algorithm runs in parallel the following two computations:

(A): Run P on v.

(B): Generate all finite strings $\Pi_1, \Pi_2, \Pi_3, \ldots$ in shortlex order and use the proof-checking algorithm to determine whether an arbitrary string Π_i is a proof of $N(P, v)$; stop when the first string Π_j is a proof for $N(P, v)$.[19]

The algorithm starts by running P on v. If $P(v)$ stops, then the computation (A) stops, the answer is "yes" and no Π_j proves $N(P, v)$ because of soundness. If $P(v)$ does not stop, then by semantical completeness, there exists a string Π_j proving $N(P, v)$; hence the computation (B) stops, and the answer is "no".

To complete the argument, we must prove that

(1) the computation "(A) and (B)" always stops,
(2) the algorithm is correct, i.e. it always gives the correct answer.

For (1), if the computation "(A) and (B)" does not stop, then "P does not stop on v" and no string Π_i proves $N(P, v)$. By semantical completeness "P does not stop on v" implies the existence of a proof Π_j for $N(P, v)$, hence the computation (B) stops, a contradiction.

For (2), we first prove that the algorithm cannot simultaneously stop on (A) and (B). Indeed, if the computation (B) stops, then a proof Π_i for

[18]In the shortlex ordering, strings are primarily sorted by length, with the shortest strings first, and strings of the same length are sorted into lexicographical order.
[19]It is possible that more strings Π_j are proofs for $N(P, v)$.

$N(P, v)$ was found, so by soundness, the computation (A) cannot stop; if, on the contrary, the computation (A) stops, then no valid proof for $N(P, v)$ can exist because otherwise, by semantic completeness, P will never halt on input v.

If the computation (A) stops, then P eventually halts on v, and the result is correct because the computation (B) does not stop. If the computation (A) does not stop, then because of semantic completeness, there will be a valid proof Π_i of $N(P, v)$ which will be eventually *discovered* by the computation (B). Having obtained the proof Π_i, we are sure (because of soundness) that P will indeed never halt.

Thus, we have described an algorithm for solving the Halting Problem, a contradiction!

Incompleteness Theorem. (Incompleteness Semantic Variant 2) *In every sound axiomatic system in which all statements of the form $N(P, v)$ are formalised and the set of proofs is decidable, there exists a true and unprovable statement $N(P_0, v_0)$, hence the system is semantically incomplete.*

First, no false statement $N(P, v)$ can be proven in the system because of the soundness. Second, the assumption that every true statement $N(P, v)$ is provable in the system contradicts the Incompleteness Semantic Variant 1.

We can easily construct an axiomatic system which decides an "unprovable" in the system true statement $N(P_0, v_0)$ by adding it as an axiom to the original system.[20] This more powerful system is still sound, hence semantic incomplete. The system is not only semantically incomplete but *not completable.* Even adding an infinity of true unprovable statements will not fix the incompatibility between soundness and semantic completeness.

We have argued that Gödel's statement G [194] as well as the statement $N(P_0, v_0)$ in ii) above are true. Is the popular formulation of the Gödel Incompleteness Theorem – *in every axiomatic system satisfying certain conditions there exist true and unprovable statements* – correct? First, defining "truth" is difficult if not impossible because of Tarski Undefinability Theorem [389],[21] see Section 6.3. However, if we have a definition of "truth" for an axiomatic system (like the one for $N(P, v)$) which is *sound,* that is, it does not permit that a statement and its negation are both true,

[20]As we noted, adding a false statement $N(P_0, v_0)$ will make the system un-sound.
[21]Informally, *arithmetical truth cannot be defined in Peano Arithmetic.*

then semantic incompleteness implies syntactic incompleteness, so the formulation above is indeed correct.[22] This problem will be discussed in more detail in Section 6.3.

There is an analogy between Poincaré's Three-Body Theorem and Gödel's Incompleteness Theorem. Both are impossibilities proven from "within" theories. Poincaré's result demolished the myth of an equational deterministic framework capable of fully predicting the evolution of the physical world, obtained by mathematical analysis followed by an original physical interpretation. Gödel's result dismantled the solution offered by Hilbert's program again by a mathematical analysis followed by a logical-philosophical interpretation. Both results have been studied intensively not only from a mathematical point of view but also from various philosophical angles.

One could argue that because both results have been obtained from "within" the theories, the negative conclusions are positive results. Indeed, we would not trust any system, organisation or device self-certifying its "good behaviour".[23]

In view of incompleteness, the job of mathematicians is not only to prove new theorems from old axioms – a task which can be relegated to a large extent to proof-assistants, see [79] and also Section 7.6 – but also to search for more and more powerful axiomatic systems; not surprisingly, Gödel's interest went more towards the second activity.

In the tension between consistency and completeness, we have chosen consistency. What about if we sacrifice consistency or a part of it? Indeed, we often have inconsistent beliefs, make inconsistent claims, or are presented with inconsistent information – inconsistency seems pervasive. Weaker than classical logics called the paraconsistent [338] attempt to model benign inconsistencies by negating the Principle of Explosion.[24] While paraconsistency cannot guarantee completeness, inconsistency opens the door to the possibility of a complete theory whose classical counterpart would be incomplete. See more in [356].

[22]See more in the section "And What About 'Truth'?" [279, p. 24].

[23]Such situations are not hypothetical as political events have shown.

[24]In Latin *ex falso [sequitur] quodlibet* (from contradiction, anything [follows]).

2.3 From the Semantic Incompleteness Theorem to the Halting Theorem

This section proves that a solution to the Halting Problem can be used to construct a sound and complete axiomatic system for proving the statements $N(P, v)$. To this aim, we consider the language of statements $N(P, v)$ as in Section 2.2 and we assume that a **halting decider** solves *correctly and in finite time the Halting Problem for every pair* (P, v). Then we construct an axiomatic system with no axioms and one inference rule:

> Run the **halting decider** on (P, v) and return YES if $P(v)$ never stops or NO otherwise.

A proof in the axiomatic system is a finite sequence of instructions executed by the **halting decider** *on input pair* (P, v)*, which ends in* YES. Explicitly, proofs in this axiomatic system are *finite sequences of instructions* $i_1, i_2, \ldots, i_n$ executed by the **halting decider** on (P, v), which ends with YES – recall that the **halting decider** stops always in finite time on any input.

If we have a proof for $N(P, v)$, then $P(v)$ does not stop by the correctness of **halting decider**, so the axiomatic system is sound, and if $P(v)$ does not stop, i.e. $N(P, v)$ is true, then there is a proof in the system for $N(P, v)$ because **halting decider** solves the Halting Problem correctly. In conclusion, the Incompleteness Theorem discussed in Section 2.2 proves the Halting Theorem.

Finally, we note that *undecidable problems*, like the Halting Problem, depend on the adopted mathematical model of computation and *unprovable statements*, like consistency, depend on a fixed axiomatic system: they are both *relative*. For example, Gödel's Incompleteness Theorem *does not show the existence of absolutely unprovable statements (in arithmetic or anywhere else), but the existence of statements which cannot be proven* in a fixed axiomatic system with some properties. Thinking about them as "absolutes" is incorrect and may lead to lunatic consequences [182].

2.4 Minimal programs

Consider

> the first natural number that cannot be named in less than fourteen words.

Why is this number so special to deserve the honour to be discussed? On one hand, as only finitely many natural numbers can be named with less than fourteen words, the number exists. On the other hand, as the above definition has thirteen words, the number does not exist. This paradox was discovered by B. Russell, who called it *Berry's paradox.*[25]

The expression "be named" in the above formulation must be made precise to be analysed mathematically. One possible way is to use programs as names for natural numbers. In this approach the *length* of a program, i.e. the number of characters of the program will be used to define the shortest programs. For example, PROGRAM(1) has 64 characters (including blanks), so its length is 64. PROGRAM(4) needs special attention as its size also depends on the value read into N, so the length of this program is bounded by the sum between $\log_t N$ – the number of characters necessary to represent N on an alphabet with t elements – and the sums of lengths of its instructions. A *(length-) minimal* [26] program is the shortest program that produces a given output, i.e. a natural number. Explicitly: a program P is *minimal* if no program shorter than P produces the same output; if more than one program satisfies the above property, then the minimal program is the first such program in shortlex order. Note that minimal programs are unique.

The following result was proved by Chaitin, see [117, Chapter 5]:

> **Minimality Theorem.** *There is no program deciding whether an arbitrary program is minimal.*

Here is a proof. Assume by the absurdity that there exists a **minimality-decider** which accepts as input a program P and returns in finite time YES or NO according to whether P is minimal or not. If **minimality-decider** returns YES on P, then P eventually stops, its output is a natural number N and no program shorter than P produces N. We recall that length(P) is the length of the program P and output(P) is the output produced by P if it stops.

Assume that the programs use t characters.

(1) For every natural N the program with "hard-coded" input N,

[25] G. G. Berry, a librarian at the Bodleian Library of Oxford University, suggested it to him; see more in [117, p. 8].
[26] Sometimes called elegant.

PROGRAM(P_N)
print N
stop

has length$(P_N) \leq \log_t N + 13$ and output$(P_N) = N$.

(2) For every natural N there exists the minimal program P_N^{min} such that output$(P_N^{min}) = N$ because the number of programs of length less than or equal to $\log_t N + 13$ producing N is finite, so one of them is minimal.

(3) With a similar argument we can show that for every natural $M > N$ there exists a minimal program whose output is M.

We next construct the program T_N, which uses the program **minimality-decider** to generate all programs P with length$(P) > N$, retains only the minimal programs and returns an output different from all outputs of the minimal programs:

PROGRAM(T_N)
generate all P with length$(P) > N$ until the program
minimality-decider finds the minimal program P_M^{min} with $M > N$
run P_M^{min}

On the one hand, for large enough N,

$$\text{length}(T_N) \leq \log_t N + \text{ constant } < N < \text{length}(P_M^{min}). \tag{2.1}$$

Indeed, as length(T_N) is the sum between the value of N, which in base t is no longer than $\log_t N + 1$, plus the sum of the lengths of instructions, we get the first inequality; the second inequality is a well-known property of the logarithm function and the last inequality comes from the construction of P_M^{min} in the program T_N.

On the other hand, the program T_N stops in finite time because of the definition of minimal programs. Furthermore, output$(T_N) =$ output$(P_M^{min}) = M$ and $T_N \neq P_M^{min}$: this contradicts (2.1).

Did the proof contradict the Halting Theorem? The Halting Theorem does not apply here because the program T_N *does not execute* the programs P (some of which do not stop), but only runs **minimality-decider** on the string representations of programs P. By definition, **minimality-decider** is a program that always stops and returns on every P, YES or NO; if **minimality-decider** returns YES on a program P, then P is guaranteed to stop, so we can safely run P.

2.5 Algorithmic complexity and incompressibility

Let us toss an unbiassed coin 100 times. According to classical probability theory, 100 heads are just as probable as any other outcome. Should we believe this?

To test our intuition, let us consider the following strings, called, in order, A, B, C, D:

0101010101010101010101010101010101010101010101010101010101010101,

0101010101010101010101010101010101010101010101010101010101010101,

1010001000011000110001010110100110100000100010010011000010001000,

0100100010111101001011110100111101001110111100111010000010001011.

Try to memorise as many strings as you can. How many did you remember? Which ones? Choose one you couldn't memorise and try to explain the reason. Which strings contain more information?[27]

One way to memorise such a string is to find a pattern and use it to reproduce the string. For example, 01 is an obvious pattern for A, which can then be reproduced by the rule: "concatenate 32 copies of 01". Can you find a pattern in B? The pattern for C is more subtle: the number of 0s is larger than the number of 1s, so memorising positions for 1s will be enough to reconstitute C.[28] The patterns discussed above allowed us to memorise a string by using a "rule" and the pattern: in other terms, all definitions have the form of a program (rule) running on an input (pattern) to produce a string. Can you find the program and input for D?

The programs we used have as input and output natural numbers. These programs can easily be transformed into *programs processing binary strings which can also be coded in binary*:[29] In this section, we will work with the pseudocode programs from Section 1.3.

[27] Challenge for the reader: why?

[28] Challenge for the reader: Does this method work for every string?

[29] One way to do it is to use a computable bijection between Σ^* and $\{0,1\}^*$, where Σ has more than two elements [61].

A *name* for a string x is a pair

$$\langle [P], w \rangle,$$

where $[P]$ is a *code*[30] of a program P and w is a string such that

$$P(w) = x,$$

that is, the computation $P(w)$ stops and produces as output x. For simplicity, we write P also for $[P]$.

The length of $\langle P, w \rangle$ is the size of the name $P(w)$. How can we define this "length"? We use a prefix-free coding: if $P = p_1 p_2 \dots p_n$, then we encode $\langle P, w \rangle$ by the string

$$p_1 0 p_2 0 \dots p_{n-1} 0 p_n 1 w. \tag{2.2}$$

We note the rule: each symbol p_i is followed by 0 except the last one, p_n, which is followed by 1, and the obtained string is concatenated with w. A simple algorithm can test whether a string x is of the form (2.2) and, in the affirmative, can compute the unique strings $p_1 p_2 \dots p_n$ and w such that $x = p_1 0 p_2 0 \dots p_{n-1} 0 p_n 1 w$. Furthermore, $\text{length}(\langle P, w \rangle) = 2\ \text{length}(P) + \text{length}(w)$.

In the spirit of Section 2.4, we consider **a** shortest (*minimal*) name $\langle M, w \rangle$ for x (several such strings may exist); its (unique) length is denoted by $K(x)$ and is called the *algorithmic complexity* indexalgorithmic complexity of x.

We can easily name a string x by placing a copy of itself directly into an analogue of PROGRAM(P_N) in Section 2.4: $P_x = x$. The length of this name is $2\ \text{length}(P) + \text{length}(x)$. This name says very little about the "structure" of x; a significantly shorter name will tell us that the information in x *can be compressed*, so its amount of information is smaller than that of x. This suggests that the algorithmic complexity is a measure of the amount of information in strings. As the PROGRAM(P_x) computes the identity function $P_x = x$, we deduce the following inequality: there exists a constant c such that for all strings x we have

$$K(x) \leq 2\ \text{length}(P) + \text{length}(x) = c + \text{length}(x), \tag{2.3}$$

[30] We can algorithmically assign binary strings to programs; see, for example, [149].

where $c = 2\,\text{length}(P)$; the inequality above follows from the minimality of K.

Formula (2.3) shows that algorithmic complexity is never much larger than the length of the string itself plus a fixed constant. Of course, the algorithmic complexity of *some* strings can be much shorter than the lengths of the strings (for example, if the information in the string appears to be redundant or sparse like in A or B). But is this true for all strings? Each name is a binary string, so the number of names of length less than n is at most the sum of the numbers of strings length $0, 1, \ldots, n-1$:

$$1 + 2 + 4 + \cdots + 2^{n-1} = 2^n - 1.$$

The number of minimal names is less than the number of strings of length n – which is 2^n, so at least one string of length n has to have $K(x) \geq n = \text{length}(x)$. We say that a string x is *c-compressible* $(1 \leq c < \text{length}(x))$ if $K(x) \leq \text{length}(x) - c$. If x is not c-compressible, then x is incompressible by c. As a corollary we show that at least $2^n - 2^{n-c+1} + 1$ strings of length n are incompressible by c. Indeed, at most $2^{n-c+1} - 1$ strings of length n are c-compressible (because at most that many descriptions of length at most $n - c$ exist), hence the remaining

$$2^n - (2^{n-c+1} - 1)$$

strings of length n are incompressible by c.

Incompressible strings, called *algorithmically random*, have many properties of randomness, i.e. the set of incompressible strings is incomputable, in a long enough incompressible string, the numbers of 0s and 1s are roughly equal and the length of its longest run of 0s is about the log of the string's length.

How robust is this model of incompressibility? To answer this question, we will show the "optimality" of the definition of the algorithmic complexity K. A *description language* is a computable function $p : \{0,1\}^* \to \{0,1\}^*$. The *minimal name* of a string x with respect to p, written $d_p(x)$, is the shortest string s in shortlex order such that $p(s) = x$ and the algorithmic complexity induced by p is

$$K_p(x) = \text{length}(d_p(x)).$$

For example, if we use Java (binary coded) as a description language, $d_{\text{Java}}(x)$ is the minimal Java program that outputs x, and $K_{\text{Java}}(x)$ is the length of the minimal program. Another way to define K is via the pseudocode programs discussed in Section 1.3: we will use it in what follows.

Universality Theorem. *For every description language p, there exists a fixed constant c, that depends only on p, such that for all strings x:*

$$K(x) \le K_p(x) + c.$$

For the proof, let p be a description language and consider the following program:

PROGRAM(P_p)
read w
$s = p(w)$
print s
stop

Then $\langle P_p, d_p(x)\rangle$ is a description of x whose length is

$$\text{length}(\langle P_p, d_p(x)\rangle) = K_p(x) + 2\ \text{length}(P_p) = K_p(x) + c,$$

where $c = 2\ \text{length}\langle P_p\rangle$.

The meaning of the Universality Theorem is the following: if a string x is incompressible by c, then x is incompressible by $c' > c$ for any other programming language (note that c' depends on the chosen programming language).

A direct consequence of the Universality Theorem is the following slightly more general version of (2.3):

Growth Theorem. *There exists a constant $c > 0$ such that for every string x we have $K(x) \le \text{length}(x) + c$.*

For proof, we consider the program $p = Id$

PROGRAM(Id)
read w
print w
stop

and we use the Universality Theorem.

Consider now $n > 0$ and $K(x)$ for all x with $\text{length}(x) = n$: the Growth Theorem shows that the complexity grows linearly on an exponential set of strings.

We can now return to the question posed at the beginning of this section. Is the classical probability theory "claim" that in tossing a fair coin, 100 heads are just as probable as any other outcome correct? To answer, we consider compressible and incompressible strings and calculate their probabilities to be drawn. The probability that a string x of length n is c-compressible is

$$\frac{\#\{x \mid \text{length}(x) = n, K(x) \leq n - c\}}{2^n} \leq (2^{n-c+1} - 1) \cdot 2^{-n} < 2^{1-c},$$

so the probability that a string x of length n is c-incompressible is

$$\#\{x \mid \text{length}(x) = n, K(x) > n - c\} \cdot 2^{-n} \geq 1 - 2^{1-c}.$$

According to classical probability theory, every string of length 100 has the same probability

$$2^{-100} = \frac{1}{1267650600228229401496703205376} \approx 7.88861 \times 10^{-31},$$

but the probability of getting a 50-compressible string of length 100 is at most

$$\frac{2^{100-50+1} - 1}{2^{100}} \approx 1.77636 \times 10^{-15},$$

hence the probability of getting a 100-bit-string which is not 50-compressible is at least

$$1 - \frac{2^{100-50+1} - 1}{2^{100}} \geq 0.99,$$

far higher than 2^{-100}. We conclude that *not all strings are equal; some contain more information than others.*

2.6 Is incompressibility decidable?

Suppose now that we choose an axiomatic system satisfying (CEN) and a variant of (CAR)[31] requiring that the statements of the form '$K(x) > m$' can be formalised in the system's language.

As in Section 2.2, we ask the question: can we choose an *axiomatic system in which all true statements '$K(x) > m$' and only them are provable?* We assume that the axiomatic is:

Sound: For all x, m, if there is a proof for '$K(x) > m$', then $K(x) > m$.
Semantically complete: For all x, m, if $K(x) > m$, then there is a proof for '$K(x) > m$'.

Next, we present two crucial results (for proofs, see Section 6.3).

How many true statements '$K(x) > m$' are provable in an axiomatic sound system? There exit infinitely many pairs (x, m) such that $K(x) > m$ because for every natural N, the set $\{x \mid K(x) \leq N\}$ has at most $1 + 2 + \cdots + 2^N = 2^{N+1} - 1$ elements. Surprisingly, only a few.

Chaitin Theorem. *Assume* (CEN) *and* Sound. *There exists a constant c (depending on the axiomatic system only) such that for every provable statement '$K(x) > m$' we have $m < c$.*

According to Chaitin Theorem, a true statement '$K(x) > m$' may be provable in the axiomatic system only if $m < c$, so only the true statements '$K(x) > 1$', '$K(x) > 2$', …,'$K(x) > c - 1$' may be provable: every statement '$K(x) > c + i$', $i = 0, 1, 2, \ldots$ is unprovable even if it is true (of course, by soundness, no false such statement is provable). Still, the set of provable theorems '$K(x) > m$' is infinite. Based on this result, we can prove the following:

Chaitin Semantic Incompleteness Theorem. (Complexity Variant) *Every axiomatic system having* (CEN) *and* (CON) *and in which statements of the form '$K(x) > m$' are formalised is semantic incomplete.*

In detail, in every axiomatic system having (CEN) and (CON) in which statements of the form '$K(x) > m$' are formalisable, there exists a true statement '$K(x) > m$' which the system cannot prove.[32]

[31] The system contains a modicum of arithmetic of natural numbers. See Section 2.
[32] In fact, infinitely many statements of the form '$K(x) > m$'.

2.7 Is algorithmic complexity computable?

As expected, the answer is negative:

> **Incomputability of K Theorem.** *The function K is not computable.*

Here is the proof. Assume for a contradiction that K is computable.

First, we note that K is unbounded, that is, for every $n > 0$, there exists a string x such that $K(x) > n$: the negation, $K(x) \leq B$, for some fixed positive integer B and all strings x, contradicts the existence of the infinity of the set of incompressible strings.

Second, fix $m \geq 0$ and define the function f from natural numbers to binary strings

$$f(m) = \min[x \mid K(x) > m],$$

where the minimum is taken according to the shortlex order. As K is unbounded, for every m there exists a string x such that $K(x) > m$, hence $f(m)$ is defined for every $m \geq 0$.

The assumption that K is computable implies that f_m is computable by a program R. The length of the program R is bounded by a constant plus the length of the binary representation of m.[33] The inequality $\log_2(m) +$ constant $< (m-2)/2$ is true for all large enough m, hence

$$\text{length}(\langle R \rangle) \leq \log_2(m) + \text{constant} < (m-2)/2,$$

for large enough m. As R does not need any input, and $\langle R, \varepsilon \rangle$ is a description of a string x with $K(x) \leq m$, we get a contradiction.[34]

The Incomputability of K Theorem poses a natural question: what relevance could have an incomputable complexity measure? The insight provided by Chaitin Theorems gives the first answer. A different answer will be shown in Section 4.4 when we will discuss the undecidability of problems in software engineering.

2.8 Coding the Halting Problem

Is it possible to code in a number all halting programs? The affirmative answer was anticipated in 1927 by the French mathematician É. Borel who

[33] Challenge for the reader: Why?

[34] Above, ε is the empty string.

"defined" the real number $0 < B < 1$ in the following way: the N-th digit b_N of B is 1 or 0 depending on whether the answer to the N-th question in an enumeration of all possible yes/no questions that one can write in French is yes or no. The number B is an oracle that "would answer all past, present, and future enigmas of science, history, and curiosity."[35]

Replacing all yes/no questions with all halting/non-halting programs, we get the first answer to the above question. Systematically enumerate all programs (say, in shortlex order) and define the real number H whose N-th bit h_N is 1 if the N-th program halts and 0 otherwise. The number H would be an extremely useful oracle for the Halting Problem to have ... but it has two drawbacks:

(1) Although there exist programs computing infinitely many bits of H, because of the undecidability of the Halting Problem, *no program can compute all the bits of* H.
(2) The oracle H for the Halting Problem is extremely wasteful, because

> N instances of the Halting Problem are not N bits of information, they are only $\log_2 N$ bits of information.

Indeed, one only needs to know *how many* of these N programs halt – a number expressed in less than or equal to $1+\log_2 N$ bits – to be able to determine which programs terminate. Just run in parallel all programs till exactly the known number of programs that halt had stopped: all the remaining programs won't halt.[36]

Using a more sophisticated version of this idea, Chaitin [114] defined the *halting probability* Ω by the following formula:

$$\Omega = \sum_{p \text{ halts}} 2^{-\text{length }(p)} < 1. \tag{2.4}$$

The number Ω can be seen as the probability that a program whose bits are generated individually by independent tosses of a fair coin will eventually halt. But is Ω an actual probability?

First, we note that there are infinitely many halting probabilities, not one because the precise numerical value of $\Omega = \Omega_L$ depends on the choice L of the programming language for the programs p.

[35]Borel argued that knowledge of every real number would have strange consequences: the number B is a case in point. This raises a troubling question: how real are real numbers? For answers see [118, 190].

[36]Challenge for the reader: Did we use this reasoning before?

Second, for the definition (2.4) to work properly, the computer programming language L the programs p are written in has to satisfy the following requirements:

- The programs p in L have to be *self-delimiting* (in L, no program is a prefix of another program).[37] This way, one can construct large programs by concatenating (abutting) routines.
- The language L has to be *concisely universal.* More precisely, for every other self-delimiting language L', there exists a constant c that depends only on L and L' such that for every program p' in L', there exists a program p in L such that $\text{output}(p) = \text{output}(p')$ (universality) and length $(p) \leq$ length $(p') + c$ (conciseness).

Such a programming language L is often referred to as a *universal self-delimiting Turing machine.* The pseudocode programs are self-delimiting (because of the syntax of the **Stop** instruction; see Section 1.3) and concisely universal. Another example is the register machine language RM, which uses the following five instructions processing information coded in registers: `=r1,r2,r3` (branching instruction), `&r1,r2` (assigning instruction), `+r1,r2` (sum), `!r1` (read one bit), and `%` (halt). Registers `r1, r2,...` can contain arbitrary large non-negative integers. A program consists of a finite list of labelled instructions from the above list, with the restriction that the halt instruction appears only once as the last instruction. The input data is a bit-string which follows immediately after the halt instruction. The result of the computation, if any, is a non-negative integer stored in a fixed output register. A program not reading the whole data or attempting to read past the last input bit results in an error.

Is the halting probability Ω interesting? Here are some arguments showing that the answer is affirmative.

By writing Ω in binary one can show that given the first N bits $\omega_1 \ldots \omega_N$ of

$$\Omega_L = \sum_{n=1,2,3,\ldots} \omega_n 2^{-n},$$

one can solve the Halting Problem for all programs up to N bits in size. This implies that there exists a fixed constant c depending only on the programming language L such that the bits of Ω are:

- *Algorithmic random*: The smallest program for computing the first N

[37]Otherwise, the sum will not converge.

bits of Ω has at least $N - c$ bits.[38]

- *Logical irreducible*: Every sound extension of **ZFC**[39] that can prove the values of the first N bits of Ω must have at least $N - c$ bits of axioms.

Given these properties, it follows that the bits of Ω are the most concise oracle for solving the Halting Problem because it gives an exponential compression of all answers to individual instances of the Halting Problem. Indeed, the first N bits of Ω allow us to answer the Halting problem for all $2^{N+1} - 1$ programs p having at most N bits in size.

The philosophical and epistemological significance of Ω is primarily due to the following fact:

> The bits $\omega_1\omega_2\omega_3 \ldots$ of the halting probability Ω provide a "perfect" simulation within pure mathematics, where all truths are necessary, of an infinite stream of contingent, accidental yes/no facts.

Every Ω provides a natural example of a *Borel absolutely normal real number*: if Ω is written in any base $b \geq 2$, then all blocks of n base-b digits will occur with equal limiting relative frequency b^{-n} [61].

Like for the oracle number H, an immediate result of Ω's algorithmic randomness and logical irreducibility is that no algorithm can compute the bits of Ω; however, in contrast with H, Ω *is strongly incomputable or bi-immune*, that is, *no algorithm can compute more than finitely many scattered bits of* Ω. Algorithmic randomness and its consequence, strong incomputability, are surprisingly mitigated by *computably enumerability*, the property that Ω is the limit of a computable sequence of rationals in the unit interval. The converse result is also true [61]:

> **Theorem.** *Every algorithmic random and computably enumerable real is an* Ω *number for some self-delimiting universal Turing machine.*

Strong incomputability implies unprovability:

> **Theorem.** *If* **ZFC** *is sound, then* **ZFC** *cannot prove more than finite values of the bits of* Ω.[40]

[38]This property is also called Martin-Löf randomness [291, 61].

[39]Zermelo-Fraenkel set theory with the Axiom of Choice.

[40]A quantum random generator based on measuring a value indefinite observable has the same property [2, 14].

Some of these bits have been computed with a combination of computation and mathematical and logical analysis. In [87], it was shown that for a "natural" choice of the self-delimiting universal Turing machine L, the first 40 bits of $\Omega = \Omega_L$[41] are

$$0001000000010000101001110111000011111010. \tag{2.5}$$

Each part can be improved in the hope of getting more exact bits. Various ideas in this sense have been discussed in [87]. For example, the analysis ignored syntactical properties of valid (register machine) programs and just assumed every prefix string could be extended to a halting program by appending 4 bits. However, on close inspection, one can see that the halt instruction may only appear after a complete (preceding) instruction has been well specified with its operator and operand arguments: this means that at most 1/4 of prefixes have their last instruction well-specified with all required operands. This fact can be used to get one more exact bit by lowering the contribution to Ω of this tail by a factor of 1/2.

Amusingly, the bits calculated in an earlier article [90] have been "used" in the CBS drama TV show Numb3rs[42] in which a mathematician who works for the FBI uses equations to help solve various crimes:

> LARRY: Ah, Charles, my ambulatory reference book. Chaitin's Omega Constant...?
>
> CHARLIE: Omega equals .00787499699. Why, what're you working on? (sees the file, reacts) Oh. FBI file.

More recently, Ω and the above computation were featured in [48].

Every number Ω includes a tremendous amount of mathematical knowledge. According to Bennett and Garnder [40]

> [Omega] embodies an enormous amount of wisdom in a very small space ... inasmuch as its first few thousand digits, which could be written on a small piece of paper, contain the answers to more

[41] Challenge for the reader: How could an Ω be "unnatural"?

[42] See `www.imdb.com/title/tt0433309/` in Season 5, episode 5, scene 6. The math is explained in `http://numb3rs.wolfram.com/505`.

> mathematical questions than could be written down in the entire universe.
>
> Throughout history mystics and philosophers have sought a compact key to universal wisdom, a finite formula or text which, when known and understood, would provide the answer to every question. The use of the Bible, the Koran and the I Ching for divination and the tradition of the secret books of Hermes Trismegistus, and the medieval Jewish Cabala exemplify this belief or hope. Such sources of universal wisdom are traditionally protected from casual use by being hard to find, hard to understand when found, and dangerous to use, tending to answer more questions and deeper ones than the searcher wishes to ask. The esoteric book is, like God, simple yet undescribable. It is omniscient, and transforms all who know it ... Omega is in many senses a cabalistic number. It can be known of, but not known, through human reason. To know it in detail, one would have to accept its uncomputable digit sequence on faith, like words of a sacred text.

If we knew 2,745 bits of Ω_{MR} we would know whether Riemann's Hypothesis is true [111]; surprisingly, the conjecture **P** versus **NP** is significantly more complex [75].[43]

Can a halting probability be formally proven algorithmically random in Peano Arithmetic (**PA**)?[44] The following facts have been proven in [93]. First, a negative result:

> **Theorem.** *There exists a concise universal self-delimiting Turing machine whose universality cannot be proven in* **PA**.

In contrast with Chaitin Theorem in Section 2.6, here is a positive result:

> **Theorem.** *If $\Omega = \Omega_L$ is given by a self-delimiting Turing machine L whose universality and conciseness are proven in* **PA**, *then* **PA** *proves that Ω is algorithmically random.*

Yet one more negative result:

> **Theorem.** *There exists a universal self-delimiting Turing machine U such that, based on U,* **PA** *cannot prove the algorithmic*

[43]See more in Section 3.9.

[44]See Section 6.1 for more about **PA**.

randomness of its halting probability.

We have seen that one important property of Ω is strong incomputability: *no algorithm can compute more than finitely many scattered bits of* Ω.[45] In fact, there exists Ω's such that *no algorithm can compute any bit of* Ω; see [69]. Still, the first 40 exact bits of a natural Ω have been computed in [87].

2.9 Algorithmic and time complexities

A program's *time complexity* ($Time$) is the number of "instructions" it takes. The time complexity of a program is either a positive integer or infinity. For example, the time complexity of PROGRAM(1) is 205, and the time complexity of PROGRAM(2) is infinite.

More interestingly, the inequality $Time(p) \leq n$ is *decidable* in contrast with the incomputability of K and the Halting Theorem.

Is there a relation between K and $Time$?

We can compute the complexity K with an oracle for deciding the Halting Problem. Indeed, if we have an oracle for deciding the Halting Problem, then we can decide, for arbitrary P, w, x, whether $P(w) = x$, then we can compute:

$$K(x) = \min\{\text{length}(\langle P, w\rangle) \mid P(w) = x\},$$

because of the inequality (2.3).

Is the converse implication true as well?

To answer this question in the affirmative, we fix an enumeration of all binary strings via the computable bijection bin between the set of positive integers and $\{0,1\}^*$, which associates to every $n \geq 1$ its binary expansion without the leading 1. For example,

The following result shows a connection between the time complexity and K:[46]

Time Lemma. [**116, 107**] *There is a constant* $c > 0$ *such that for every program* p *that stops in time* t *the inequality* $K(\text{bin}(t)) \leq \text{length}(p)) + c$ *holds true.*

[45] This property is called bi-immunity [84].

[46] More results will be discussed in Section 5.11.

n	n_2	bin(n)	length(bin(n))
1	1	λ	0
2	10	0	1
3	11	1	1
4	100	00	2
$\vdots$	$\vdots$	$\vdots$	$\vdots$

Table 2.1: The computable bijection bin

The above inequality depends on the length(p), not p. This means that the bound works for all programs of a given length.

Assume now that K is computable by an oracle. Following [120], we can decide the Halting Problem using K. Given a program p, we compute the largest T such that $K(\text{bin}(T)) \leq \text{length}(p)) + c$ and run p for the time T: if the computation does not halt, then the program p never stops.[47]

[47]Challenge for the reader: Fix $i > 0$. Is the set $\{x \mid K(x) \leq i\}$ infinite?

Chapter 3

Famous Mathematical Statements and the Halting Problem

Consider the following mathematical statements: Euclid's Theorem of the infinity of primes, Goldbach's Conjecture, Fermat's Last Theorem, Hilbert's Tenth Problem, the Four Colour Theorem, Riemann's Hypothesis, the Collatz Conjecture, and the **P** versus **NP** Problem. What do they have in common? Very little, except their notoriety. Some are (famous) conjectures; some are well-known theorems. They belong to different areas of mathematics, number theory, Diophantine equations, complex analysis, graph theory, and complexity theory. Assertions state positive facts or negative results; some are very old, and others are relatively new.

Surprisingly enough, such a "commonality" exists for many mathematical problems, and one of them – which will be discussed in this section – is based on the possibility of expressing the problem in terms of (very) simple programs reducible to the Halting Problem. As a consequence, a uniform approach for evaluating the complexity of a large class of mathematical problems was developed [77, 72, 74].

In this chapter, we search for a "common computational structure" in some interesting mathematical problems, including those enumerated above.

3.1 The infinity of primes

Euclid is credited with the first proof that the set of primes is infinite: there is no largest prime as there is no largest natural number. The typical argument runs as follows. Let us suppose that the set of primes is finite, say $\{p_1, p_2, p_3, \ldots, p_n\}$. Construct the number $q = p_1 p_2 p_3 \cdots p_n + 1$. If q is a prime, then we have a new prime as $q > p_i$ for all $1 \leq i \leq n$. If q is not a prime, then (by the Fundamental Theorem of Arithmetic[1]) it must be divisible by a prime $r < q$. But r cannot be any p_i in our original exhaustive list of primes because dividing q by p_i produces the remainder 1. As a consequence, r is a new prime. In both cases, we have found a prime not in the original list, a contradiction.

Wittgenstein[2] criticised Euclid's proof for its "totality" nature because:

> In mathematics, we must always deal with systems, not with totalities.

Instead of a proof of the existence of an infinity of primes based on deduction from certain formal or informal assumptions, Wittgenstein called for the construction of "a formal expression that proves the infinity of primes by its syntactical features".[3]

Such proof can be easily obtained by rephrasing Euclid's argument in terms of a program Π_{primes}, which generates the sequence of primes in increasing order: the infinity of primes is equivalent with the non-halting of the program Π_{primes}.

3.2 The Pigeonhole Principle

The statement

> If $n > 0$ pigeons are put into $m < n$ pigeonholes, there is a pigeonhole with more than one pigeon.

is called the Pigeonhole Principle or the Dirichlet Principle.

A formal statement is:

> Every function from a set of n elements into a set with $m < n$ elements is not injective.

A program $\Pi_{\text{pigeonholeprinciple}}$ which for all n generates all functions from $\{1, 2, \ldots, n\}$ into $\{1, 2, \ldots, m\}$, for all $m < n$, and checks for each of them whether the function is injective will either find an injective function and

[1] Every integer greater than one can be represented uniquely as a product of prime numbers, up to the order of the factors.

[2] Who advocated a form of anti-platonism by rejecting the interpretation of mathematical propositions in terms of propositions that are capable of being true or false in correspondence to reality [261, p. 64]

[3] Although we do not subscribe to Wittgenstein's philosophy of mathematics, we acknowledge the importance of his preference for process vs. function.

stop, or will continue forever. The validity of the Pigeonhole Principle is equivalent to the fact that the program $\Pi_{\text{pigeonholeprinciple}}$ never stops.

3.3 Goldbach's Conjecture and other problems in number theory

This conjecture,[4] included in Hilbert's eighth problem [221], is

> **Goldbach's Conjecture.**[5] All positive even integers greater than two can be expressed as the sum of two primes.

The conjecture was tested up to $400,000,000,000,000$.[6]

Applying the same idea as for the infinity of primes, one can write a (computer) program Π_{Goldbach}, which enumerates all positive even integers greater than two, and for each of them, checks the required property; the program Π_{Goldbach} stops if and only if it finds a counter-example for Goldbach's conjecture. Re-phrased: Π_{Goldbach} never stops if and only if Goldbach's conjecture is true.

The same approach works for many problems in number theory. One example is:

> **Fermat's Last Theorem.**[7] *There are no natural numbers $n > 2, x, y, z$ such that $x^n + y^n = z^n$.*

Indeed, the program Π_{Fermat} systematically generates all 4-tuples of positive integers greater than 3, (n, x, y, z), and stops when it finds the first 4-tuple for which $x^n + y^n = z^n$.

Another example is

> **Mersenne's Conjecture.**[8] There are infinitely many prime numbers of the form $2^n - 1$ (Mersenne numbers).

---

[4]Proposed on 7 June 1742 by the German mathematician C. Goldbach in a letter to L. Euler,`tinyurl.com/2p9e3sm8`.

[5]`tinyurl.com/2p9e3sm8`.

[6]`towardsdatascience.com/prime-numbers-and-goldbach-s-conjecture-visualization-60d1993a1424`; note also the visualisation.

[7]In 1637, P. de Fermat wrote in his copy of the *Arithmetica by Diophantus of Alexandria* (c. 250 CE), "It is impossible for a cube to be a sum of two cubes, a fourth power to be a sum of two fourth powers, or in general for any number that is a power greater than the second to be the sum of two like powers. I have discovered a truly remarkable proof, but this margin is too small to contain it."

[8]`en.wikipedia.org/wiki/Twin_prime#Twin_prime_conjecture`.

A prime p is a twin prime if $p+2$ is also prime. For example, 41 is a twin prime because 43 is also prime. Sometimes the term twin prime is used for a pair of twin primes. Twin primes become increasingly rare, keeping with the general tendency of gaps between adjacent primes to increase as the numbers get larger.

Still, another example is

> **The Twin Prime Conjecture.**[9] There exist infinitely many twin primes.

It is clear that Goldbach's conjecture and Fermat's Last Theorem are statements of the form:

$$(\forall n)\, P(n), \tag{3.1}$$

where P is a computable predicate. The last two conjectures have a more complicated structure, i.e. they can be written in the form

$$(\forall N)\, (\exists n > N)\, P'(n),$$

where P' is a computable predicate. A program generating more and more natural numbers satisfying the Twin Prime Conjecture may not stop because there are infinitely many pairs of primes $p, p+2$ or because there are only finitely many primes p such that $p+2$ is also prime!

The halting property of an associated program cannot represent the last two conjectures. It is an *open question* whether the conjectures can be described in the form (3.1). Still, can they be described regarding the halting status of programs?

3.4 Hilbert's Tenth Problem

Solving algebraic equations using integer (or rational) constants in (positive) integers is an old mathematical activity. Some of these equations do not have solutions at all; others have finitely many solutions or infinitely many solutions. The equation $2x-2y = 1$ has no integer solutions, the equation $5x = 10$ has a unique integer solution, while the equation $7x - 17y = 1$ has infinitely many integer solutions.[10]

[9] Proposed in 1846 by the French mathematician A. de Polignac.
[10] Challenge for the reader: Are these statements correct?

A Diophantine equation[11] is an equation of the form $P = 0$ where P is a polynomial with integer coefficients. For example, Fermat's equations $x^n + y^n = z^n$, for $n = 1, 2, \ldots$ are all Diophantine. To solve a given Diophantine equation $P = 0$, one must determine whether the equation has solutions in (positive) integers and, if it has, to find all of them.

The original formulation of Hilbert's Tenth Problem is:[12]

> **10.** Determining the solvability of a Diophantine equation. Given a Diophantine equation with any number of unknowns and with rational integer coefficients: devise a process, which could determine by a finite number of operations whether the equation is solvable in rational integers.

Consider a parametric Diophantine equation:

$$P(a_1, a_2, \ldots, a_n, x_1, x_2, \ldots, x_{m+1}) = 0, \tag{3.2}$$

where $a_1, a_2, \ldots, a_n$ are parameters and $x_1, x_2, \ldots, x_{m+1}$ are unknowns. Fixing values for parameters results in a particular Diophantine equation. For example, in the parametric Diophantine equation $(a_1 - a_2)^2 - x_1 - 1 = 0$, the parameters are a_1, a_2 and x_1 is the only unknown. If we put $a_1 = 1, a_2 = 0$ we get the Diophantine equation $x_1 = 0$; if we take $a_1 = a_2 = 0$ we get the Diophantine equation $x_1 + 1 = 0$.

Given a parametric Diophantine equation (3.2), one can construct a program Π_P which, beginning with the input $a_1, a_2, \ldots, a_n$, will eventually halt if and only if equation (3.2) has a solution in the unknowns $x_1, x_2, \ldots, x_{m+1}$. The program Π_P systematically generates all vectors with $m+1$ integers $(i_1, i_2, \ldots, i_{m+1})$, checks for each of them whether $P(a_1, a_2, \ldots, a_n, i_1, i_2, \ldots, i_{m+1}) = 0$ and stops when the first solution is found.

Furthermore, we can construct a program Π_H10P which, given an arbitrary Diophantine equation (without parameters, i.e. $n = 0$) $P = 0$, eventually stops if and only if the equation $P = 0$ has a solution.

Can we decide whether the program Π_H10P eventually halts in a finite time? It is well-known that the answer is negative [153, 293]. This result, called the MRDP Theorem (an initialism for the surnames of the four principal contributors to its solution – Y. Matiyasevich, J. Robinson, M. Davis, and H. Putnam), needed 21 years to be proven; Matiya-

[11] Named after Diophantus of Alexandria (III century AD).

[12] For the original statement in German see [220].

sevich completed the last step in 1970.[13] The core argument [153] is based on the fact that every computably enumerable set of natural numbers can be represented in the form $\{n \in \mathbf{N} \mid P(n, x_1, x_2, \ldots, x_m) = 0$ has a solution in the non-negative integers unknown $x_1, x_2, \ldots, x_m\}$.

Fig. 3.1: Y. Matiyasevich[14]

It is interesting to note that the counterpart of Hilbert's Tenth Problem for *real* unknowns, that is, given an equation of the form $P(x_1, x_2, \ldots, x_m) = 0$, where P is a polynomial with integer coefficients (like in the classical case), but $x_1, x_2, \ldots, x_m$ are real unknowns, is decidable: there is a program which decides in a finite amount of time whether the equation has a solution in the domain of reals. The decision problem is solved by Sturm method for $m = 1$; Tarski's method [390] works for any number of unknowns.[15] Of course, no program can, in general, compute exactly some solutions, even if their number is known, because solutions can be irrational; however, solutions can be effectively approximated up to any precision. This leads us to the following problem for standard Diophantine equations.

Fix a Diophantine equation:

$$D(n, x_1, x_2, \ldots, x_m) = 0, \tag{3.3}$$

and consider the following two questions:

[13]The result was presented by the 22-year-old Y. Matiyasevich in his Doctoral Thesis at the Leningrad Department of the Steklov Institute of Mathematics.
[14]Source: Y. Matiyasevich, email to C. Calude on 4 June 2023.
[15]This shows that extrapolating computational results from positive integers to reals are not always possible.

(1) for a fixed $n = n_0$, does the equation $D(n_0, x_1, x_2, \ldots, x_m) = 0$ have a solution?
(2) for a fixed $n = n_0$, does the equation $D(n_0, x_1, x_2, \ldots, x_m) = 0$ have an infinity of solutions?

Both questions are undecidable for *some* equations (3.3). For *each* equation (3.3) the information contained in the sequence of k yes/no answers to the questions "does the equation $D(n_0, x_1, x_2, \ldots, x_m) = 0$ have a solution, for $n = 1, 2, \ldots, k$?" contains only $\log k$ bits of information (knowing how many equations have solutions is enough to determine exactly which equations have solutions).[16] This information can be algorithmically compressed. However, for *some* equations (3.3), the information contained in the sequence of k yes/no answers to the question "does the equation $D(n_0, x_1, x_2, \ldots, x_m) = 0$ have infinitely many solutions, for $n = 1, 2, \ldots, k$?" contains about k bits of information, i.e. this information cannot be algorithmically compressed. See more in [61].

3.5 The Four Colour Theorem

In graph-theoretical terms, the Four Colour Theorem states that the vertices of every planar graph can be coloured with at most four colours so that no two adjacent vertices receive the same colour. Shortly, every planar graph is four-colourable.

The theorem was proven in 1977 [26, 25] (see also [410]) using a computer-assisted proof which consists in constructing a finite set of "configurations" and a proof that each of them is "reducible", which implies that no configuration with this property can appear in a minimal counter-example to the theorem. Checking the correctness of the original proof was a challenging task: it implies, among other things, checking the input descriptions of 1476, later 633, graphs, checking the correctness of the programs, proving the correctness of the compiler used to compile the programs, and checking the reliability of the hardware used to run the programs.[17] Various partial independent verifications have been obtained,[18] culminating with the formal proof in [341].

A program $\Pi_{\text{fourcolourtheorem}}$, which systematically generates all planar

[16]Challenge for the reader: Is there any connection with the Decidability Criterion in Section 1.8?

[17]This computer-assisted proof generated many mathematical and philosophical discussions about the acceptability of a mathematical proof; see for example [59, 79, 80].

[18]There is no verification in its entirety.

graphs and checks for each of them whether it is colourable with four colours and stops when the first counter-example is found will never halt if and only if the theorem is true. However, this program will be quite long because testing the planarity of a graph is difficult. In [153] a better solution was proposed by constructing first a Diophantine equation

$$F(n,t,a,\ldots)=0, \tag{3.4}$$

which has no solution if and only if every planar graph can be coloured with at most four colours so that no two adjacent vertices receive the same colour.

A significant improvement can be obtained by observing that without restricting the generality we can consider only the maps T_n consisting of the points (x,y) such that $J(x,y) \leq Q = (n^2+3n)/2$, where J is Cantor's bijection $J(x,y) = ((x+y)^2+3x+y)/2$. Given a 4-colouring of T_n, $t_0, t_1, \ldots, t_Q$ there exist (and can be effectively computed) s, t such that for all $0 \leq i \leq Q$:[19]

$$t_i = rem(t, 1+s(i+1)).$$

Accordingly, the sequence $t_0, t_1, \ldots, t_Q$ can be coded by s and t: every sequence $u_0, u_1, \ldots, u_Q$ with $u_i \leq 44$ can be represented by some $u \leq R = (1+4(Q+2)!)^{Q+1}$ such that

$$u_i = rem(u, 1+4(Q+2)!(i+1)).$$

Finally, there is a map (say T_n) which cannot be coloured in 4 colours if and only if the following condition is satisfied:

$$(\exists n,t,s)(\forall u \leq R)(\exists x,y)(x+y \leq n)[A \vee B],$$

where

$$A = u_{J(x,y)} \geq 4,$$

$$\begin{aligned} B = [&(t_{J(x,y)} = t_{J(x+1,y)} \wedge u_{J(x,y)} \neq u_{J(x+1,y)}) \\ &\vee (t_{J(x,y)} \neq t_{J(x+1,y)} \wedge u_{J(x,y)} = u_{J(x+1,y)}) \\ &\vee (t_{J(x,y)} = t_{J(x,y+1)} \wedge u_{J(x,y)} \neq u_{J(x,y+1)}) \\ &\vee (t_{J(x,y)} \neq t_{J(x,y+1)} \wedge u_{J(x,y)} = u_{J(x,y+1)})]. \end{aligned}$$

A simple inspection shows that the above condition is decidable, so the Four Colour Theorem is of the form $(\forall n)P(n)$, where P is a computable predicate.

[19] The integer remainder function is denoted by *rem*.

3.6 Riemann Hypothesis

Riemann Hypothesis is probably the most famous/important conjecture in mathematics. It appears in Hilbert's eighth problem [221]: the non-trivial complex zeros of Riemann's zeta function, which is defined for $Re(s) > 1$[20] by:

$$\zeta(s) = \sum_{n=1}^{\infty} \frac{1}{n^s},$$

lie exactly on the line $Re(s) = 1/2$, a surprisingly interesting regularity. This hypothesis has been checked for the first 10,000,000,000,000 solutions, but no proof is yet known. The first proof of Riemann's Hypothesis will be rewarded with $\$1,000,000$ by the Clay Mathematics Institute.[21] A proof of Riemann's Hypothesis would shed light on many of the mysteries surrounding the distribution of prime numbers because Riemann showed that the frequency of prime numbers is very closely related to the behaviour of the function ζ.

According to Matiyasevich [293, pp. 119–121], the negation of Riemann's Hypothesis is equivalent to the existence of positive integers k, l, m, n satisfying the conjunctions of the following six conditions:

(1) $n \geq 600$,
(2) $\forall y < n\, [(y+1) \mid m]$,
(3) $m > 0\, \&\, \forall y < m\, [y = m \vee \exists x < n\, [\neg\, [(x+1) \mid y]]]$,
(4) $explog(m-1, l)$,
(5) $explog(n-1, k)$,
(6) $(l-n)^2 > 4n^2k^4$,

where $x \mid z$ means "x divides z" and $explog(a, b)$ denotes the predicate

$$\exists x\, \left(x > b + 1\, \&\, (1 + 1/x)^{xb} \leq a + 1 < 4(1 + 1/x)^{xb}]\right).$$

An inspection of the above conditions shows that Riemann's Hypothesis is of the form $\forall n, R(n)$, where R is a computable predicate. Hence, one can write a program Π_{Riemann} such that Riemann's Hypothesis is false if and only if Π_{Riemann} halts.

[20]The real part of a complex number s.
[21]`www.claymath.org/millennium-problems/riemann-hypothesis`.

3.7 Euler's Integer Partition Theorem

The number of ways of writing an integer as a sum of n positive integers, where the order of addends is ignored, is denoted by $P(n)$. By $P(n|\text{odd parts})$ and $P(n|\text{distinct parts})$, we denote the number of ways of writing the integer as a sum of n odd/distinct positive integers. By convention, partitions are usually ordered from largest to smallest. Some examples are presented in Table 3.1.

n	odd parts	$P(n\|\text{odd parts})$	distinct parts	$P(n\|\text{distinct parts})$
1	1	1	1	1
2	$1+1$	1	2	1
3	$1+1+1$		3	
	3	2	$2+1$	2
4	$1+1+1+1$		4	
	$3+1$	2	$3+1$	2
5	$1+1+1+1+1$		5	
	$3+1+1$		$4+1$	
	5	3	$3+2$	3
6	$1+1+1+1+1+1$		6	
	$3+1+1+1$		$5+1$	
	$3+3$		$4+2$	
	$5+1$	4	$3+2+1$	4

Table 3.1: Integer partitions into odd integers vs. integer partitions into distinct integers

L. Euler is credited (cf. [186, p. 2]) with the proof in 1748 of the theorem bearing his name: no matter how long we extend Table 3.1, there will always be as many items in the left column as in the right one. In other words, *the number of partitions of an integer into odd integers equals the number of partitions into distinct integers.* In [81] it was proved that Euler's Integer Partition Theorem has the form $(\forall n)P(n)$, where P is a computable predicate, and its complexity – which is incomputable – was evaluated.

3.8 Collatz's Conjecture

L. Collatz posed the following problem in 1937, two years after receiving his doctorate:[22]

[22]Also known as Collatz's conjecture, the Syracuse conjecture, the $3x + 1$ problem, Kakutani's problem, Hasse algorithm, or Ulam's problem.

Given any integer seed a_1 there exists a natural N such that $a_N = 1$, where

$$a_{n+1} = \begin{cases} a_n/2, & \text{if } a_n \text{ is even}, \\ 3a_n + 1, & \text{otherwise}. \end{cases}$$

There is a vast literature on this problem and various natural generalisations; a good presentation is [260].

P. Erdös said (cf. [209, pp. 330–336]) that

> Mathematics may not be ready for such problems.

Many mathematicians explored experimental and heuristic arguments in support of its truth. The conjecture has been checked for values up to $2^{68} \approx 2.95 \times 10^{20}$.[23] Other methods include probabilistic heuristics, the use of proof-assistants, and the analysis of stopping. Interesting visualisations of the stopping times are presented in the Wikipedia entry for this conjecture.[24] Recently T. Tao [387] proved one of the most significant results on this problem:

> **Tao Theorem.** *Almost all orbits of the Collatz map attain almost bounded values.*

Does there exist a program Π_{Collatz} such that Collatz's Conjecture is false if and only if Π_{Collatz} halts?

First, we note that a brute-force tester, i.e. the program which will enumerate all seeds, and for each of them, will try to find an iteration equal to 1, can never stop in two different cases: a) because the conjecture is indeed true, b) because for some specific seed a_1, there is no N such that $a_N = 1$. It is not clear how to differentiate these cases; even worse, it is not clear how to refute b) by brute-force testing.

Here is a simple non-constructive argument answering in the affirmative this question [75]. Indeed, observe first that the set

$$Collatz = \{a_1 \mid a_N = 1, \text{ for some } N \geq 1\}$$

is computably enumerable. Collatz's Conjecture requires proving that *Collatz* contains all positive integers.

If *Collatz* is not computable, then the conjecture is false, and Π_{Collatz} can be taken every program which eventually halts as a) is ruled out.

[23] As of 27 July 2023.

[24] See `en.wikipedia.org/wiki/Collatz_conjecture`.

If *Collatz* is computable, then we can write a program Π_{Collatz} to find an integer not in *Collatz*:[25] the conjecture is true if and only if Π_{Collatz} never stops. Furthermore, Π_{Collatz} is of the form $\forall x R(x)$, where R is a computable predicate. If the statement "Π_{Collatz} halts" is independent of **ZFC** – neither the statement nor its negation is provable in **ZFC** – then *the Collatz Conjecture is true.* Indeed, if the Collatz Conjecture would be false, then Π_{Collatz} will stop in a finite time, hence **ZFC** would prove "Π_{Collatz} halts", a contradiction. A proof of independence is an alternative proof – admittedly not usual – of the Collatz Conjecture. The above conclusion is true not only for Π_{Collatz}, but for every program of the form $\forall x R(x)$, where R is a computable predicate.

Which of the above decision programs Π halt? The program Π_{Fermat} never stops as certified by A. Wiles' proof of Fermat's Last Theorem. The answer is not known for the program Π_{Riemann}. The question is open to all conjectures/hypotheses discussed in this section. We [currently] cannot even write the program explicitly Π_{Collatz}.

3.9 The P versus NP Problem

The **P** versus **NP** Problem asks whether every problem whose solutions can be quickly verified can also be quickly solved.

The class of decision problems solvable by quick/feasible algorithms is called **P**. The class of questions for which an answer can be quickly verified is denoted by **NP**.

An algorithm (solving, verifying) is "considered" to be *quick* or *feasible* if it runs in polynomial-time on the size of the input. Obviously, an algorithm solving a problem in exponential time is not quick, but is an algorithm running in polynomial-time quick? More precisely, is the polynomial-time algorithm the "correct", "adequate" mathematical model for feasible computation? An affirmative answer is provided by Cobham's Thesis.[26] How credible is Cobham's Thesis? Is it as "credible" as the Church-Turing Thesis discussed in Section 1.12? According to Davis [236, pp. 568–569], the answer is negative.[27] In [82], it was conjectured that there is no good

[25] Writing Π_{Collatz} does not need the computability of *Collatz*.

[26] `www.cs.toronto.edu/~sacook/homepage/cobham_intrinsic.pdf`.

[27] In the discussions following J. Hartmanis' invited lecture *Turing Machine Inspired Computer Science Results*, CiE2012, 22 June 2012, `http://www.mathcomp.leeds.ac.uk/turing2012/WScie12/Content/abstracts/juris.html`, M. Davis asked the question he posed the speaker about 30 years ago: "How would you feel if **P** = **NP** with a polynomial of degree 100?" Hartmanis' original answer was: "God cannot be so cruel!".

mathematical model for feasible computation.

Fig. 3.2: J. Hartmanis[28]

The class **NP** includes many pattern-matching and optimisation problems that are of great practical interest, such as determining the optimal arrangement of transistors on a silicon chip, developing accurate financial-forecasting models, analysing protein-folding behaviour in a cell, and the QUBO problem, which is essential for quantum annealing and D-Wave; see Section 1.13.[29]

Suppose you have to organise housing accommodations for 400 university students, but only one-quarter can be accommodated in the dormitory. Additionally, you have been given an enumeration of pairs of "incompatible" students, and you must ensure that no pair from this list appears in the final list of 100 students. This is an **NP** problem because it is easy to check if a proposed list of 100 students is acceptable. However, generating such a list from scratch "seems" impractical due to the enormous number of possible combinations. We said "seems" because whether there is a *quick* solution is unknown. This issue falls under the **P** versus **NP** Problem.

[28]Source: In Wikipedia. `en.wikipedia.org/wiki/Juris_Hartmanis`.

[29]Sudoku puzzles and airline-schedule searches on `bing.com` are also in **NP**.

Fig. 3.3: S. Cook[30]

This problem, formulated by Cook [135] in 1971 and independently by Levin [271] in 1973, is considered one of the most challenging open problems in theoretical computer science/mathematics.

Fig. 3.4: L. A. Levin[31]

[30]Source: In Wikipedia. www.wikiwand.com/en/File:Prof.Cook.jpg.

[31]Source: In Wikipedia. www.wikiwand.com/en/File:LeonidLevin2010.jpg.

The episode TV-14 (aired 3 October 2013) in the show *Elementary*,[32] a modern retelling of Sherlock and Watson's adventures, deals with the killing of Felix Soto, a mathematics professor who was trying to solve the **P** versus **NP** Problem. Were Soto's equations worth killing over?

3.10 Finitely refutable statements

The Clay Mathematics Institute will award a prize of one million dollars for its first correct solution.[33] A constructive proof for **P**=**NP** based on an efficient simulation would have significant practical consequences, for example, in cryptography, which relies on certain problems being difficult to solve; a proof for $\mathbf{P} \neq \mathbf{NP}$ (which is widely believed to be the case) would lack practical computational benefits, but would have important theoretical value.

A problem is *finitely solvable* if it can be solved by examining finitely many cases.

For example, consider the predicate:

$$P(n) = \begin{cases} 1, & \text{if } n \text{ is even or } n = 1 \text{ or } n \text{ is a prime}, \\ 0, & \text{otherwise}, \end{cases} \tag{3.5}$$

that is, $P(n) = 0$ if and only if n is an odd, not prime number greater than 1.

The predicate P defined in (3.5) is computable. Is the problem $(\forall n)(P(n) = 1)$ decidable? The brute-force approach of deciding $P(1) = 1?$, $P(2) = 1?, \ldots, P(n) = 1?, \ldots$ could continue for ever. However, the predicate $(\forall n)\,(P(n) = 1)$ is *finitely solvable*, hence it is sufficient to check all n up to 9 to refute it. Indeed, $\neg((\forall n)\,(P(n) = 1))$ if and only if $(\exists n)(n$ odd, not prime number greater than 1) and $n = 9$.

More generally, consider the formula

$$f = Q_1 n_1\, Q_2 n_2\, \ldots\, Q_k n_k\, P(n_1, n_2, \ldots, n_k),$$

where $Q_1, Q_2, \ldots, Q_k \in \{\forall, \exists\}$ are quantifier symbols. We say that f is in the class $\hat{\Pi}_s$ or $\hat{\Sigma}_s$ if the quantifier prefix of f starts with $\forall$ or $\exists$, respectively, and contains $s - 1$ alternations of quantifier symbols. When P is computable, then f is in Π_s or Σ_s, respectively. It is sufficient to consider only such formulæ f in which no two consecutive quantifier symbols are the same; in the sequel, we make this assumption without special mention.

[32]`https://www.imdb.com/title/tt3125780/`.

[33]`www.claymath.org/sites/default/files/pvsnp.pdf`.

As usual, with P as above, we write $P(n_1, \ldots, n_s)$ instead of $P(n_1, \ldots, n_s) = 1$; $n_1, \ldots, n_s \in \mathbf{N}$. Thus, $\neg P(n_1, \ldots, n_s)$ if and only if $P(n_1, \ldots, n_s) = 0$. Moreover, since we consider variable symbols only in the domain $\mathbf{N}$, if f is any formula in first-order logic, we write f is true instead of f is true in $\mathbf{N}$.

Let Γ_s be one of the classes $\hat{\Pi}_s$, $\hat{\Sigma}_s$, Π_s, and Σ_s. We refer to the task of proving or refuting a first-order logic formula as a *problem* and for problems expressed by formulæ in Γ_s as Γ_s*-problems.*

We say that a problem is being *solved* if the corresponding formula is proven or disproved, that is, if the truth value of the formula is determined.

What is the "commonality" of all problems in classes $\hat{\Pi}_s$ and $\hat{\Sigma}_s$?

Following [62], we define a *test set* for f to be a set $T \subseteq \mathbf{N}^s$ such that f is true in $\mathbf{N}^s$ if and only if it is true in T. The problem f is *finitely solvable* if there is a finite test set for f. We have:

Finite Solvable Theorem. [62] *Every $f \in \hat{\Gamma}_s$ is finitely solvable.*

In other words, a solution to each $f \in \hat{\Gamma}_s$ can be obtained by inspecting only finitely many instances of the problem.

The statement of the Finite Solvable Theorem has some similarities with the *Test Set Theorem* in formal language theory:[34]

Test Set Theorem. [68] *Let X and Y be alphabets, and let $L \subseteq X^*$. There exists a finite subset F of L – the test set – such that for every two morphisms f, g from X^* to Y^*, $f(u) = g(u)$ for all $u \in L$ whenever $f(u) = g(u)$ for all $u \in F$.*

Recall that a function $f\colon X^* \to Y^*$ is a morphism if $f(\varepsilon) = \varepsilon$ and for every strings $x, y \in X^*$, $f(xy) = f(x)f(y)$. As proven in [68], the existence of the test sets is not constructive.[35] In the statement of the Test Set Theorem, the order of the quantifiers, that is, $\forall L \exists F \forall f \forall g$, is very important: modifying the order $\forall L \forall f \forall g \exists F$ produces a far weaker statement.

As we might expect, there is no uniform algorithmic way to solve finitely solvable problems because the finite test sets cannot be computed even for all problems in the class Π_1:

---

[34] Originally known as Ehrenfeucht's Conjecture.

[35] Under special assumptions on L like regularity, test sets can be effectively constructed.

Non-Constructivity Theorem. [62] *There is no constructive proof showing that every $f \in \Pi_1$ has a finite test set.*

Even determining an upper bound for $f \in \Pi_1$ is as incomputable as the Halting Problem [62].

The *Busy Beaver Function* $\Sigma : \mathbf{N} \to \mathbf{N}$ [333] gives an upper bound on the number of operations executed by a halting Turing machine of a given size. Formally, it is defined as follows:

$$\Sigma(n) = \max\{\text{length}(U(x)) \mid \text{length}(x) = n \text{ and } U(x) \text{ halts}\}.$$

As a result, the Busy Beaver Function[36] is incomputable [333]. In fact, Σ grows faster than any computable function.

Consider P a computable unary predicate on $\mathbf{N}$ and $f = \forall n\, P(n)$, hence $f \in \Pi_1$. Consider a program p_f for U:

$$U(p_f) = \begin{cases} \min\{n \mid \neg P(n)\}, & \text{if } f \text{ is not true,} \\ \infty, & \text{otherwise.} \end{cases}$$

Such a program[37] always exists. For example, the program which computes $P(1)$, $P(2)$, ... and halts when the first n satisfies $\neg P(n)$.

Let $m_f = \text{length}(p_f)$. If f is not true, then U halts on p_f and the output is $\nu(f) \leq \Sigma(\text{length}(p_f)) = \Sigma(m_f)$. If f is true, $U(p_f) = \infty$ and then $\nu(f) = 0$. This proves the following statement:

Upper Bound Theorem. [62] *For every $f \in \Pi_1$, $\nu(f) \leq \Sigma(m_f)$.*

To decide the truth of f, one only needs to check the truth value of $P(n)$ for all n not exceeding $\Sigma(m_f)$. This could be very useful if Σ were computable, but this is not true as we mentioned before. Hence, the bound $\nu(f) \leq \Sigma(m_f)$ does not help in the actual solution of the problem of f.

The above solution fails, but, of course, this does not mean that there is *no solution*. In fact, there is a solution which follows from the fact that a *no computable bound exists!* Here is the argument. For any $\pi \in X^*$, define the predicate P_π on $\mathbf{N}$ by

$$P_\pi(n) = \begin{cases} 1, & \text{if } U(\pi) \text{ does not halt within } n \text{ steps,} \\ 0, & \text{otherwise.} \end{cases}$$

[36] Its name comes from the fact that Σ computes so quickly and prolifically, like a beaver running back and forth.

[37] Challenge for the reader: Are there more such programs?

Clearly, the predicate is computable.[38] Define $f_\pi = \forall n\, P_\pi(n)$ and notice that f_π is true if and only if $U(\pi)$ does not halt.

Assume now that there is a program to compute an upper bound of $\nu(f)$ for any $f \in \Pi_1$; this program takes, as input, a program ρ computing the predicate P^ρ whose output is a natural number $\nu'(\rho)$ such that $\nu(f^\rho) \leq \nu'(\rho)$, where $f^\rho = \forall n\; P^\rho(n)$. We next prove that this assumption implies the existence of an algorithm deciding the Halting Problem (for Turing machines). Indeed, consider $\pi \in X^*$. To decide whether $U(\pi)$ halts, first compute a program p_π computing P_π. Next compute $\nu'(p_\pi)$. As $f_\pi = f^{p_\pi}$, one has $\nu(f_\pi) \leq \nu'(p_\pi)$. Hence, to determine whether f_π is true, it is sufficient to determine whether $P_\pi(n)$ for all $n \leq \nu'(p_\pi)$. If so, then $U(\pi)$ halts; otherwise, it doesn't. This ends the proof of the Non-Constructivity Theorem.

3.11 Evaluating the complexity of a class of mathematical statements

Evaluating (or even guessing) the degree of difficulty of an open or even a solved problem is notoriously hard not only for beginners, but also for the most experienced mathematicians.

Is there a (uniform) method to evaluate, in some objective way, the difficulty of a mathematical statement or problem? The question is not trivial because mathematical problems can be so diverse: the revised Mathematics Subject Classification (MSC2000),[39] based on two databases, Mathematical Reviews and Zentralblatt MATH, contains 63 two-digit classifications, 529 three-digit classifications, and 6,022 five-digit classifications. How could one compare a problem in number theory with a problem in complex analysis or a problem in algebraic topology? Is there any indication that all, most, or even a large part of mathematical problems have a kind of "commonality" allowing a uniform evaluation of their complexity?

Surprisingly enough, such "commonalities" do exist for many mathematical problems. A mathematical "common" property of all statements discussed in the above sections is their "logical structure": all can be represented as Π_1-statements. This is obvious for some statements, like Goldbach's Conjecture and Fermat's Last Theorem; this is not trivial to prove for others, like Riemann's Hypothesis. Collatz's conjecture is a special case: the only proof that its statement is Π_1 is non-constructive, that is,

[38] Challenge for the reader: Why?

[39] `mathscinet.ams.org/msnhtml/msc2020.pdf`.

the corresponding predicate $P(n)$ is shown to exist, but it is not explicitly constructed.[40]

As a consequence, uniform approaches for evaluating the complexity of a large class of mathematical problems/conjectures/statements can be and have been developed. See [72, 74] for overviews.

3.12 A universal prefix-free binary Turing machine

We briefly describe the syntax and the semantics of a register machine language which implements a (natural) minimal universal prefix-free binary Turing machine U; see [74].

Any register program (machine) uses a finite number of registers, each of which may contain an arbitrarily large non-negative integer.

By default, all registers, named with a lower or upper case letter string, are initialised to 0. Instructions are labelled by default with $0, 1, 2, \ldots$

The register machine instructions are listed below. In all cases, R2 and R3 denote either a register or a non-negative integer, while R1 must be a register. When referring to R, we use, depending upon the context, either the name of register R or the non-negative integer stored in R.

=R1,R2,R3

If the contents of R1 and R2 are equal, then the execution continues at the R3-th instruction of the program. If the contents of R1 and R2 are not equal, execution continues with the next instruction in sequence. If the content of R3 is outside the program's scope, then we have an illegal branch error.

&R1,R2

The content of register R1 is replaced by R2.

+R1,R2

The content of register R1 is replaced by the sum of the contents of R1 and R2.

!R1

One bit is read into the register R1, so the contents of R1 becomes either 0 or 1. Any attempt to read past the last data-bit results in a run-time error.

[40]There is no known proof showing that a constructive proof does not exist.

%

This is the last instruction for each register machine program before the input data. It halts the execution in two possible states: successfully halt or halt with an under-read error.

A *register machine program* consists of a finite list of labelled instructions from the above list, with the restriction that the halt instruction appears only once as the last instruction of the list. The input data (a binary string) follows immediately after the halt instruction. A program not reading the whole data or attempting to read past the last data-bit results in a run-time error. Some programs (as the ones presented in this paper) have no input data; these programs cannot halt with an under-read error.

The instruction `=R,R,n` is used for the unconditional jump to the n-th instruction of the program. We use integers 0 = `false` and 1 = `true` for Boolean data types.

For longer programs, it is convenient to distinguish between the main program and some sets of instructions called "routines" which perform specific tasks for another routine or the main program. The call and call-back of a routine are executed with unconditional jumps.

3.13 Binary coding of register machine programs

In this section, we develop a systematic, efficient method to uniquely code the register machine programs in binary. To this aim, we use the following prefix-free coding.

The binary coding of special characters (instructions and comma) is in Table 3.2 (ε is the empty string).

special characters	code	special characters	code
,	ε	+	111
&	01	!	110
=	00	%	100

Table 3.2: Special characters

For registers we use the prefix-free code $\text{code}_1 = \{0^{\text{length}(x)}1x \mid x \in \{0,1\}^*\}$. The codes of the first 32 registers are in Table 3.3.[41]

[41]The register names are chosen to optimise the length of the program, i.e. the most

register	code_1	register	code_1
R_1	010	R_{17}	000010010
R_2	011	R_{18}	000010011
R_3	00100	R_{19}	000010100
R_4	00101	R_{20}	000010101
R_5	00110	R_{21}	000010110
R_6	00111	R_{22}	000010111
R_7	0001000	R_{23}	000011000
R_8	0001001	R_{24}	000011001
R_9	0001010	R_{25}	000011010
R_{10}	0001011	R_{26}	000011011
R_{11}	0001100	R_{27}	000011100
R_{12}	0001101	R_{28}	000011101
R_{13}	0001110	R_{29}	000011110
R_{14}	0001111	R_{30}	000011111
R_{15}	000010000	R_{31}	00000100000
R_{16}	000010001	R_{32}	00000100001

Table 3.3: Registers

For non-negative integers we use the prefix-free code $\text{code}_2 = \{1^{\text{length}(x)}0x \mid x \in \{0,1\}^*\}$. The codes of the first 16 non-negative integers are displayed in the following Table 3.4.

The instructions are coded by self-delimiting binary strings as follows:

(1) `& R1,R2` is coded in two different ways depending on R2:[42]

$$01\text{code}_1(\text{R1})\text{code}_i(\text{R2}),$$

where $i = 1$ if R2 is a register and $i = 2$ if R2 is an integer.

(2) `+ R1,R2` is coded in two different ways depending on R2:

$$111\text{code}_1(\text{R1})\text{code}_i(\text{R2}),$$

where $i = 1$ if R2 is a register and $i = 2$ if R2 is an integer.

(3) `= R1,R2,R3` is coded in four different ways depending on the data types of R2 and R3:

---

frequent registers have the smallest code_1 length.

[42]As $x\varepsilon = \varepsilon x = x$, for every string $x \in \{0,1\}^*$, in what follows we omit ε.

integer	code_2	integer	code_2
0	100	8	1110010
1	101	9	1110011
2	11000	10	1110100
3	11001	11	1110101
4	11010	12	1110110
5	11011	13	1110111
6	1110000	14	111100000
7	1110001	15	111100001

Table 3.4: Non-negative integers

$$00\text{code}_1(\text{R1})\text{code}_i(\text{R2})\text{code}_j(\text{R3}),$$

where $i = 1$ if R2 is a register and $i = 2$ if R2 is an integer, $j = 1$ if R3 is a register and $j = 2$ if R3 is an integer.

(4) `!R1` is coded by

$$110\text{code}_1(\text{R1}).$$

(5) `%` is coded by

$$100.$$

All codings for instruction names, special symbol comma, registers and non-negative integers are self-delimiting; the prefix-free codes used for registers and non-negative integers are disjoint. The code of any instruction is the concatenation of the codes of the instruction name and the codes (in order) of its components; hence the set of codes of instructions is prefix-free. The code of a program is the concatenation of the codes of its instructions, so the set of codes of all programs is prefix-free too.

The smallest program which halts is 100, and the smallest program which never halts is 00010010100100.

3.14 Programming techniques

An essential tool for coding sequences is Cantor's bijection maps (codes) a pair of non-negative integers a, b into a single non-negative integer $\langle a, b\rangle = 1+2+\cdots+(a+b)+a$. This function can be iterated to a bijection between $\mathbf{N}^k$

and **N**, for every $k > 1$; we shall adopt, by convention, the left-associative iteration. For example, to work with arrays in register machine programs, we need to code (finite) sequences of non-negative integers into a single non-negative integer.

In what follows, we use Cantor's bijection for such codings. For example the 4-element sequence $[2, 1, 1, 0]$ is encoded by 1484 as $\langle\langle\langle 2, 1\rangle, 1\rangle, 0\rangle = \langle 8, 1\rangle, 0\rangle = \langle 53, 0\rangle = 1484$. The reverse process allows converting of, for each given $k \geq 1$, any non-negative integer into a unique k-element sequence of non-negative integers. For example, the number 5564 can be converted to the 4-element sequence $[3, 1, 0, 0]$ and the 2-element sequence $[104, 0]$.

We can program Cantor's bijection using the elegant mathematical formula $\langle a, b\rangle = (a + b)(a + b + 1)/2 + a$[43] or with a shorter program, using the mathematically "ugly" formula $\langle a, b\rangle = 1 + 2 + \cdots + (a + b) + a$; see Table 3.5.

nr	instruction	code	length
0	`&e, a`	01 011 010	8
1	`+e, b`	100 011 00110	11
2	`&d, 0`	01 00101 100	10
3	`=e, 0, 9`	101 011 100 1110011	16
4	`&f, 1`	01 00100 101	10
5	`+d, f`	100 00101 00100	13
6	`=e, f, 9`	101 011 00100 1110011	18
7	`+f, 1`	100 00100 101	11
8	`=a, a, 5`	101 010 010 11011	14
9	`+d, a`	100 00101 010	11
10	`=a, a, c`	101 010 010 00111	14

Table 3.5: Cantor's bijection with 11 instructions

This register machine program has 136 bits:

01011010100011001100100101100101011100111001101001001011000010100
10010101100100111001110000100101101010010110111000010101010101001
000111

[43]Challenge for the reader: Try it.

Some programs are complex, so we sometimes present parts of them as algorithms in pseudo-code to facilitate understanding. The notation used in these algorithms is self-explanatory (e.g. the assignment instruction is denoted by `Set x to v`, `Next x` is the successor, `GoTo Ln` specifies the unconditional jump, etc.).

We start with an example of routine REM that computes the integer remainder of a divided by b. A local register z is initialised with b and incremented by 1 until it reaches the value of a when the algorithm finishes. The value of d, the result of the algorithm, is initialised to 0 and incremented every time z is incremented. When d reaches the value of b, the value of d is reset to 0.

```
Algorithm REM
INPUT: a>=b>=1
OUTPUT: rem(a,b) in d, i.e  a=b*q+d, with 0<=d<b, for some q
01. Set z to b
02. Set d to 0
03. if z = a
04.    then STOP return d
05.    else Next z
06.         Next d
07.         if d = b
08.            then Set d to 0 //reset the remainder to 0
09.                 GoTo L03
10.            else GoTo L03
```

The register machine program corresponding to REM is

```
//REM  returns integer remainder of a divided by b, assumes a>=b>1
//5 instrs x4 chars=20, 4 instr x6 chars=24 chars, total 9 instrs,
//44 chars
LREM: &e,b
      &d,0
LR2:  =e,a,c  //STOP
      +e,1
      +d,1
      =d,b,LR1
      =a,a,LR2
LR1:  &d,0
      =a,a,LR2
```

We continue with two algorithms. The algorithm MUL performs the multiplication of a and b and stores the product in d. The algorithm is based on the formula $a+a+a+\ldots+a$ (b times)$= a \times b$. The local counter, h keeps track of how many times a was added.

```
Algorithm MUL
INPUT: a>=0, b>=0
OUTPUT: a*b in d
01. Set d to 0
02. if b = 0
03.    then return d,STOP
04     else Set h to 1
05.         Set d to a
06.         if h = b
07.            then return d,STOP
08.            else Next h
09.                 Set d to d+a
10.                 GoTo  L06
```

The algorithm CMP returns 0 if its two input values a and b are equal, returns 1 when $a < b$ and returns 2 when $b > a$.

```
Algorithm CMP
INPUT: a>=0,b>=0
OUTPUT: d is 1 if a<b, d is 0 for a=b and d is 2 otherwise
01.Set e to a
02.Set f to b
03.Next e
04.Next f
05.Set d to 0
06.if e=f
07.    then STOP
08.    else Set d to 1
09.         if  e= b
10.     then STOP
11.     else Set d to 2
12.                 if f=a
13.                    then STOP
14.                    else GoTo L03
```

3.15 A complexity measure for Π_1-statements

The complexity measure [163, 72, 74] for Π_1-statements is defined using register machine programs which implement a universal self-delimiting Turing machine U. The machine U, as described in Section 3.12, has to be *minimal* in the sense that none of its instructions can be simulated by a program for

U written with the other instructions.

To every Π_1-problem $\pi = \forall m P(m)$, we associate the algorithm $\Pi_P = \inf\{n \mid P(n) = \text{ false}\}$, which systematically searches for a counter-example for π. There are many programs (for U) that implement Π_P; without loss of generality, any such program will also be denoted by Π_P. Note that π is true iff $U(\Pi_P)$ never halts.

The complexity (with respect to U) of a Π_1-problem π is defined by the length of the smallest-length program (for U) Π_P – defined as above – where minimisation is calculated for all possible representations of π as $\pi = \forall n P(n)$ with P a computable predicate:[44]

$$C_U(\pi) = \min\{\text{length}(\Pi_P) \mid \pi = \forall n P(n)\}.$$

Because the complexity C_U is incomputable,[45] we work with upper bounds for C_U. As the exact value of C_U is not important, following [74] we classify Π_1-problems into the following classes:

$$\mathfrak{C}_{U,n} = \{\pi : \pi \text{ is a } \Pi_1\text{-problem}, C_U(\pi) \leq n \text{ kbit}^{46}\}.$$

3.16 Concrete complexities

In this section, we present programs for Goldbach's Conjecture and Riemann's Hypothesis whose complexities are respectively in $\mathfrak{C}_{U,1}$ and $\mathfrak{C}_{U,3}$; see [111, 74].

We first present the program Π_{Goldbach}:

```
      =a,a,L0  //jump to start Goldbach program
//SUB routine returns a-b assumes a>=b>=1
//Total:  6 instrs, 4 of 4 chars= 16,
//2 of 6 chars=12, Total:  28 chars
LSUB: &x, b
      &d,0
LS1:  =x,a,c
      +x,1
      +d,1
      =a,a LS1
//REM routine returns integer remainder
//of a divided by b in d, a>=b>=2
//Total:  9 instrs, 4 of 6 chars =24,
```

[44]For C_U it is irrelevant whether π is known to be true or false. In particular, the program containing the single instruction halt is not a Π_P program for any P.
[45]Challenge for the reader: Why?
[46]A kilobit (kbit or kb) is equal to 2^{10} bits.

```
//5 of 4 chars=20 chars, Total: 44 chars
LREM: &x, b
       &y, 0
LR4: =x,a,c
     +x,1
     +y,1
     =y, b, LR1
     =a,a,LR4
LR1: &y,0
     =a,a,LR4
//IS PRIME ROUTINE: IPR, routine returns 1 if a>=2 prime,
//otherwise returns 0, calls REM
//Total:18 instructs, 13 of 4 chars=52 chars
// 4 of 6 chars=24chars and 1 of 2 chars=2, Total: 78 chars.
LIPR:&aa,a
     &bb,b,
     &cc,c
     &p,2
     &d,1
LI1: =a,p, LI5
     &b, p    // call REM a,p
     &c, LI3
     #REM
     &q,d
LI3: =q,0, LI4
     +p,1
     =a,a,LI1
LI4: &d,0
LI5: &a,aa  //prepare exit
 &b,bb
 &c,cc
     =a,a, c
//---------------Goldbach starts here
L0: &e, 4          //enumeration of all even numbers >=4
L1: &f, 2         //enumeration of all primes >=2
L2: =f, e, L8
    &c, L5       //calling IS-PRIME(a), with a = f= NUM
    &a, f
    #IPR    //returned from IS-PRIME
    &g, d
L5: =g, 1, L4
L3: +f, 1     // NUM( or f) not prime, try next NUM
    =f, f, L2      // jump to L2
L4: &c, L6        // calling a - b (named SUB)
    &a, e
    &b, f
```

```
    #SUB    //returned from SUB
    &g, d          // h is now e-f (or EVEN-NUM)
L6: &c, L7      //calling IS_PRIME(a), with a = h= EVEN-NUM
    &a, g
    #IPR    //returned from IS-PRIME
    &g, d
L7: =g, 0, L3
    +e, 2  //test next EVEN
    =a, a, L1
L8: &d, 0
    %
```

The "binary" form of the program used to calculate its size is the following:

```
0000  =a,a,10000
0001  & e,10
0010  & d,1
0011  = a,e,c
0100  & d,0
0101  & f,e
0110  = f,a,1101
0111  + f,1
1000  + d,1
1001  = d,e,1011
1010  = a,a,0110
1011  & d,0
1100  = a,a,0110
1101  = d,0,c
1110  + e,1
1111  = a,a,0010
0001 0000 & g,100
0001 0001 & h,10
0001 0010 = g,h,100110
0001 0011 & c,10110
0001 0100 & a,h
0001 0101 # 0001
0001 0110 = d,0,100011
0001 0111 & i,0
0001 1000 & k,h
0001 1001 = k,q,11101
0001 1010 + i,1
0001 1011 + k,1
0001 1100 = a,a,1101
0001 1101 & c,100000
0001 1110 & a,i
```

```
0001 1111 # 0001
0010 0000 = d,0,100011
0010 0001 + g,10
0010 0010 = a,a,10001
0010 0011 + h,1
0010 0100 = a,a,10010
0010 0101 & d,0
0010 0110 %
```

The program has 57 instructions and 247 characters (6 instructions and 28 chars from SUB, 9 instructions and 44 characters from REM, 18 instructions and 78 characters from IPR, 24 instructions and 97 characters from the main program, 20 characters from labels). So the total number of characters is 267, which gives a total of 1,068 bits.

The program for Riemann Hypothesis is more complex:

```
&a,Riemann
=b,c,RIEM     // b=c=0 so jumps to start of Riemann algorithm
// CMP(a,b) returns 1 if a<b, 0 if a=b, or 2 if a>b
//It does not invoke any other routine
//7 instr x 4 chars= 28 chars,4 instr x 6 chars=24 chars;
//11 instr, Total:  52 chars
LCMP:&e,a
&f,b
LC1: +e,1
+f,1
&d,0
=e,f,c
&d,1
=e,b,c
&d,2
=f,a,c
=a,a,LC1
// Pow(a,b) returns a^b where we assume a>0 and b>0
//It uses  MUL(a,b) =a*b--a  will not change, so it is not copied
//on the local stack
//9 instr x 4 char = 6 chars, 2 instr x 6=12 chars, 1 instr of 2
//chars=2
//Total:12 instrs = 50 chars
LPOW:&e,b     // keep our runtime stack for b,c
&f,c
&x,1
&d,a
LP1:=x,b,c
&b,d  // get ready to compute a*d
```

```
&c,LP3
#MUL    // replace with the address of routine MUL
LP3: +x,1
&b,e       // unpop stack to restore b and c
&c,f
=a,a,LP1
//SUB routine returns a-b assumes a>=b>=1
//it does not call any other routine
//4 instr x 4 chars=16 chars, 3 instructs x 6 chars= 18,
//Total:  7 instrs,  34 chars
LSUB: &x, b
      &d,0
LS1:  =x,a, LS2
      +x,1
      +d,1
      =a,a LS1
LS2:  =a,a,c
//REM routine returns integer remainder of a divided by b
// and assumes a>=b>2
//it does not call any other routine
//5 instrs x4 chars=20, 5 instr x6 chars=30 chars,
//Total: 10 instructions, 50 chars

LREM: &x, b
      &d, 0
LR4: =x,a, LR3
     +x,1
     +d,1
     =d, b, LR1
     =a,a,LR4
LR1: &d,0
     =a,a,LR4
LR3: =a,a,c

//MUL returns in d the product of a and b-to add here
//7 instructs, 4 of 4 chars= 16 chars;
//3 of 6 chars=18 chars, total 34 chars
LMUL: &d,0
    =b,0,c      //stop and return d
    &x,1
LM1:+d, a
   = x,b,c      //stop and return d
    +x,1
    =a,a LM1
//------------------------------------------
// Explog(a+1,b); uses global variable ah as upper bound
```

```
// to test if there exists ah >= x > b+1 such that
// (x+1)^{xb} <= (a+1)x^{xb} < 4(x+1)^{xb}
// It uses: MUL, POW, and CMP.
//41 instrs 4 chars=164, 6  instrs 6 chars= 36,
// 7 instrs of 2 chars=14 chars
//Total:  54 instructs,  214chars
&d,0
=ah,0,c     // return if no upper bound set
&ba,a       // store parameters locally
&bb,b
&bc,c
&bd,b       // start checking x at b+2
+bd,2
L1: &a,bd  // compute x+1 and x*b and store into be and bf
&be,bd
+be,1
&a,bd
&b,bb
&c,L2
#MUL
L2: &bf,d
&a,be       // compute bg = (x+1)^{bf}
&b,bf
&c,L3
#POW
L3: &bg,d
&a,bd       // compute x^{bf} and this times a+1, yielding bh
&c,L4
#POW
L4: &a,d
&b,ba
&c,L5
#MUL
L5: &bh,d
&a,bg       // test first <=
&b,bh
&c,L6
#CMP
L6: &a,L9
=d,2,a      // if false try next x
&a,4        // else compute 4*bg
&b,bg
&c,L7
#MUL
L7: &a,bh  // test second <
&b,d
```

```
&c,L8
#CMP
L8: &a,L10
=d,1,a     // test if found x
L9: +bd,1  // next x
&d,0
&a,L10
=bd,ah,a   // exit if reach upper bound
&d,L1
=a,a,d
L10: &a,ba // return, so restore initial values
&b,bb
&c,bc
=a,a,c
//==========================================

// Riemann uses: SUB, MOD, EXPLOG,MUL,CMP
//It has 98 lines of instructions
//It has 78 instrs of 6 chars= 468 chars,
// 10 instr of 4 chars=40 chars
// 9 instrs of 2 chars =18 chars
//and 1 instr of 1 char,
// Total: 527 chars

&ca,599    // start of Riemann conjecture testing program
L1: +ca,1
&cb,600
L2: &ah,0
L3: &cc,0
L4: &cd,0
L5: &b,cb  // compute ce=ca-cb-ah-cc-cd (=ca-b)
+b,ah
+b,cc
+b,cd
&a,ca
&c,L6
#SUB    //-----------------use of SUB
L6: &ce,d
&cf,0      // check Cond 1
L9: +cf,1  // next y
=cf,cb,L7   // did we check all y<n (if yes, check Cond 2)
&a,cf
&b,cd
&c,L8
#MOD //-----------------use of MOD
L8:=d,0,L9    // divides when a mod b is zero
```

```
=a,a,L10
L7: &cf,1  // check Cond 2
LA2: +cf,1      // next y
=cf,cd,LA1   // did we check all y<m (if yes, check Cond 3)
&be,0
=cd,0,L10    // check m>0 (kludge!)
LA4: =be,n,L10
+be,1       // did we try all x < n?  (next m, if so)
&a,be
&b,cf
&c,LA3
#MOD
LA3: &c,LA4
=d,0,c      // divides so try next x
=a,a,LA2     // do next y
LA1: &a,cd // check Cond 3
&b,cc
&c,LA5
#EXPLOG
LA5: &c,L10
=d,0,c      // if fail do next m
&a,cb       // check Cond 4
&b,ce
&c,LA6
#EXPLOG
LA6: &c,L10
=d,0,c      // if fail do next m
&a,cb       // check Cond 5
&b,ce
&c,LA9
#MUL
LA9: &bg,d
+bg,bg      // bg=2nk
&a,cc
&b,cb
&c,LA7
#CMP
LA7: &c,LA8 // if l > n
=d,2,c
&a,cb        // else l <= n
&b,bg
+b,cc
&c,LB0
#CMP
LB0: &c,LX
=d,2,c      // last condition passes
```

```
&c,L10     // else try next m
LA8: &a,cc
&b,bg
+b,cb
&c,LB1
#CMP
LB1: &c,LX
=d,2,c      // 2nd way last condition passes
L10: &bg,cd // next m
+bg,cb
+bg,ah
+bg,cc
=bg,ca,L11   // check at end of m loop
+cd,1      // else increment
&d,L5
L11: &bg,cc // next l
+bg,cb
+bg,ah
=bg,ca,L12  // check at end of l loop
+cc,1
=a,a,L4
L12: &bg,ah // next b
+bg,cb
=bg,ca,L3  // check at end of b loop
+ah,1
=a,a,L3
L13: +cb,1
=cb,ca,L1 // next n and next s
=a,a,L2
LX: %
```

The program consists of 96 instructions for routines, 93 instructions in the main program giving a total of 1,087 characters (including 196 characters from labels), hence a total length of 4,348 bits.

Fermat's Last Theorem is in $\mathfrak{C}_{U,1}$ [74], the Four Colour Theorem is in $\mathfrak{C}_{U,4}$ [73], and (surprisingly) the Euler's Integer Partition Theorem is in $\mathfrak{C}_{U,7}$ [81].[47] Related results can be found in [83]. The complexity of Riemann's Hypothesis is close to the complexity of the Four Colour Problem (4,336) [73] and is almost four times higher than the complexity of Fermat's Last Theorem (1,388) [74]. Improving the size of the programs for these statements is still possible. However, because the technique was used uniformly, it is unlikely that the complexity of the Four Colour Theorem can

[47]Mainly because of the codification of arrays.

be decreased to a value comparable with the complexity of Fermat's Last Theorem. However, the relationship between the complexities of the Four Colour Theorem and Riemann Hypothesis is more delicate.

Finally, can the method presented in this chapter compare the two open problems in the Clay Mathematics Institute list, Riemann Hypothesis and the **P** versus **NP** Problem? The main difficulty is that it is not known whether **P** versus **NP** Problem is a Π_1-statement. In [82], an inductive computation was used to find a counter-example. With the new inductive complexity measure, the **P** versus **NP** Problem *is considerably more complex than* Riemann Hypothesis.

Chapter 4

Sagacious Applications of the Halting Theorem

4.1 Can software testing and validating be done by computers only?

Did you notice the never-ending updates every non-trivial app sends to its customers? Does this mean that we can never trust software? How can we rely on a program to control spacecraft or critical medical equipment?

Sure, we can test the software, so what is the problem? Software testing is essential. Updates – based on results produced by testing – may seem free, but they are not. The cost of testing software roughly equals that of software development [289]. Why is it so expensive? Testing is a complex process whose cost depends on answers to many questions, including the following:

(1) How much time should be spent on testing?
(2) How many dedicated testing resources should be used?
(3) What levels of software testing are reasonable?

Why should we choose "reasonable testing" and not "extensive testing" or, even better, "true testing"? Why not write a (complex) testing program which automatically tests and validates every software, ideally by *proving* the correctness of the result?

E. W. Dijkstra, one of the most influential members of computing science's founding generation, said [162]:

> Program testing can be used to show the presence of bugs but never to show their absence![1]

[1] I have checked the files of this book several times. Does it mean that the book is bug-free?

First, to show the absence of bugs in a program, we have to prove that the program *halts* – can the program (suddenly?) loop forever? Testing whether a program loops forever is algorithmically decidable, see Section 1.6, but the Halting Problem is undecidable! This is the ultimate reason testing is so essential, complex, expensive and never guarantees correctness, so it never ends. In [407], the decidability of several problems related to optimisation and program testing is investigated: in particular, a detailed proof of the undecidability of testing is given.

Modern software testing involves the evaluation of one or more properties of interest to evaluate to what extent they are satisfied. For example, does the system

(1) meet the requirements that guided its design and development,
(2) respond correctly to various inputs,
(3) perform its functions within an acceptable time,
(4) run in its intended environment,
(5) achieve the expected results.

The three most popular testing techniques are called boxes:

- Black-box tests are performed to detect bugs and determine whether software corresponds to functional requirements without knowledge of the system's internal code.
- In white-box testing, the internal logic is closely examined, and tests of internal structures or workings are performed.
- Gray-box is a combination of white-box testing and black-box testing to search for improper structure or usage of applications.

There are various standards, like IEEE and ISO 9000 standards, which are periodically updated, partly caused by the use of concepts with different meanings. Partial automatic testing has been intensively studied. One method consists in the generation of automatic goal-oriented decidable test cases [289]; other tools use machine learning. The February 2023 issue of *New Scientist* includes a note titled "ChatGPT can find and fix the bugs in computer code" [380]. ChatGPT[2] was compared with other AI-powered tools used to detect and fix program bugs [369]. The paper evaluated ChatGPT on the standard bug-fixing benchmark set QuixBugs, and its performance was compared with the results of several other approaches. The analysis concluded that ChatGPT's bug-fixing performance

[2]The Chatbot developed by the company OpenAI, `openai.com`.

is competitive with the standard deep learning approaches CoCoNut and Codex and "notably better than the results reported for the standard program repair approaches". In contrast to all other approaches, ChatGPT accepts further information, for example, the expected output for a particular input or an observed error message. Such hints helped ChatGPT to record a higher success rate, namely fixing 31 out of 40 bugs: this result outperforms the current state-of-the-art.

4.2 Can we write an anti-virus program which never needs any update?

How many times do we see anti-virus update notifications? Are we not tired of them and, more importantly, of their costs? Cybersecurity experts advise constantly updating the anti-virus software because new viruses regularly threaten computers, and there is no better way to combat them. The signature files provided by anti-virus software containing the latest lists of known viruses are released periodically (sometimes daily, and even more often).[3]

Anti-virus programs are complex, expensive and can cause problems. A "false alarm" is triggered when the anti-virus software identifies a non-malicious file as malware. These cases are not frequent, but they can cause serious problems when they happen. For example, if an anti-virus application is configured to immediately delete or quarantine infected files, which is not uncommon, an essential file falsely identified as malicious can render the operating system or part of it unusable, with severe consequences.

So, why not write an anti-virus program which never needs any update? To answer this question, we need to formalise the concept of "virus". Once we have such a definition, some basic questions can be formulated and, hopefully, answered:

(1) How hard is it to detect programs infected by computer viruses?
(2) Can infected programs be 'disinfected'?
(3) What forms of protection exist?
(4) How destructive can computer viruses be?

[3]In mid-September 2021, Apple, which has a reputation for producing the most secure and safe operating systems, had to issue an emergency software patch to block the so-called "zero-click" spyware that could infect iPhones and iPads [429]. A flaw which lets hackers access devices through the iMessage service even if users do not click on a link or file was identified by independent researchers.

A possible solution is to use the notion of a self-reproducing program: J. von Neumann [405] showed mathematically that a program could reproduce itself.

Fig. 4.1: J. von Neumann[4]

There are many examples of such programs: quine[5] is a computer program which takes no input and produces a copy of its source code as its only output. The standard strategy to write a quine is to have, within the program, two pieces: (a) a code used to do the actual printing and (b) the data that represents the textual form of the code. Below is a small example in Python 3.8.[6]

```
print((_:='print((_:=%r)%%_)')%_)
```

F. Cohen[7] wrote in 1988 a Doctoral dissertation [129] on computer viruses in which he confirmed experimentally von Neumann's mathematical result and investigated other properties of viruses, such as detectability and self-obfuscation. He illustrated von Neumann's result with the following `contradictory-virus` example that cannot be detected by a virus detector D:

[4]Source: In Wikipedia. `en.wikipedia.org/wiki/File:JohnvonNeumann-LosAlamos.gif`.

[5]The name was coined by D. Hofstadter in his popular science book [227] in honour of logician W. Quine (1908–2000), `en.wikipedia.org/wiki/Willard_Van_Orman_Quine`, who extensively studied self-reference.

[6]Author: M. Weinberger, September 2020, `stackoverflow.com/users/8291579/marcus-weinberger`.

[7]Cohen is also known for the invention of high-integrity operating system mechanisms, and automation of protection management functions.

```
program contradictory-virus :=
{....
     main-program :=
     {if ~D(contradictory-virus) then
        {infect-executable;
        if trigger-pulled then
            do-damage;
         }
     goto next;
     }
 }
```

As Cohen noted (cf. [353]), "if the decision procedure D determines CV to be a virus, CV will not infect other programs, and thus will not act as a virus. If D determines that CV is not a virus, CV will infect other programs and thus be a virus. Therefore, the hypothetical decision procedure D is self-contradictory, and precise determination of a virus by its appearance is undecidable".

Cohen's PhD advisor, L. Adleman, rigorously proved that, in the general case, algorithmic determination of the presence of a virus is undecidable [12]. Adleman's proof is considered one, if not the deepest, result about malware proved in the framework of computability theory and, not surprisingly, relies on the Halting Problem's undecidability. Ironically, Young and Yung [419] proved that Adleman's work could be used to construct a virus that is highly resistant to reverse-engineering using the notion of cryptovirus.[8]

Cohen's paper [130] contains a formalisation of sets of transitive integrity-corrupting processes called *viral sets* and proves that *viruses are at least as computationally powerful as Turing machines*, hence they are affected by the undecidability of the Halting Problem.

A review of undecidable problems in this area, including virus detection, detecting unpacking execution, matching malware samples against a set of given templates and detecting trigger-based behaviour, can be found in [353]. The book [357] is an entertaining account of hacking through

[8]A cryptovirus is a computer virus that contains and uses a public key. For example, a virus or worm may generate and use its key pair at run-time. Cryptoviruses utilise secret sharing to hide information and may communicate by reading/writing posts from public bulletin boards.

examples written by a law professor and computer geek. Most interestingly, the book argues the importance of human behaviour as a critical part of hacking.

4.3 Can we write a perfectly secure operating system?

According to G. Spaffword:[9]

> The only truly secure system is one that is powered off, cast in a block of concrete and sealed in a lead-lined room with armed guards – and even then I have my doubts.

The doubt is not surprising after reading the previous section. Still, with all the high-tech protection offered by long and short-lived passwords, good anti-virus software, one-time code generators, SMS message codes, disk encryption, etc. why don't we have not a perfectly secure operating system, but at least a highly secure operating system? The standard answer is that everything is flawed. A more detailed answer involves complexity, fallibility, probability, and ... opinion [416]. Computers operate in a hybrid echo system, including themselves and humans, all interacting in myriad ways. Computers run on programs written ultimately by humans (i.e. even, the automatically generated codes are produced by human-written programs), which are fallible[10] like the various components in complex hardware (manufacturing defects, premature failures, even dust). However, from careless operations to downright malicious behaviour, human behaviour is probably the biggest threat to security. The best strategy is to almost collectively strive to make this echo system *as secure as possible.*

4.4 Are there undecidable problems in software engineering?

According to F. Brooks Jr.[11]

> More software projects have gone awry for lack of calendar time than for all other causes combined. Why is this cause of disaster so common?

[9] `en.wikiquote.org/wiki/Gene_Spafford`.

[10] It is generally agreed that human codes include about 15–50 errors per 1,000 lines, about half being potential security vulnerabilities.

[11] `courses.cs.vt.edu/~cs1104/HLL/Brooks.html`.

This fact is acknowledged across the domain. Various software-based methods have been designed to predict development times. For example, software cost models estimate development time as a function of a size measure of the product. However, the estimations could be more precise.

Why? Here is an illustration, not an answer. Assume that your friend has just enrolled into an online piano programme[12] and you have been asked to estimate how long she would need to finish the programme with a given level. Before answering, you will most likely ask a few questions: How much knowledge do you have about music theory? Did you play the piano before? If yes, what is your competence? How much time are you going to spend practising every day? She will answer some questions, but, again, quite likely, for others, her answer will be "I do not really know, but please give me a time estimate." It is not difficult to estimate (no pun!) the accuracy of such an estimate.

Can a "size measure" determine or estimate *a priori* the development time with a feasible algorithm, that is, an algorithm that returns the output in a reasonable amount of time assuming average human and computer capabilities?

We have shown in Section 2.7 that the algorithmic complexity K is incomputable. But is K relevant for modelling the complexity of "real-world programs"? This question was first posed and answered in the affirmative by Lewis [273], and the conclusion is:

> ... algorithmic complexity results can be directly interpreted as indicating that software development schedules, productivity, and reliability cannot be objectively and feasibly estimated and so will remain a matter of intuition.

Here is a sketch of Lewis' arguments.

First, we connect algorithmic complexity and "real-world programs". Although the definition of the complexity K primarily focuses on the output, it can also be used for interactivity, input and state.

(1) Interactivity: An interactive program can be seen as a series of subprograms invoked sequentially based on user commands. These subprograms share subroutines and maintain a global state. Temporarily disregarding input and state, the program's complexity consists of the complexity of the combined subprograms, along with the complexity of

[12]Like Hoffman Academy, `www.hoffmanacademy.com` or J. See, `www.jeremysee.com`.

the small event loop responsible for invoking the subprograms according to user input.
(2) Input: The complexity of a function that relies on input arguments can be defined as the complexity of a large table that contains the input-output pairs interleaved in a specific manner. Additionally, it includes the complexity of a scheme used to separate the input-output pairs and the complexity of a small program that processes the input arguments and produces the desired output.
(3) State: A state can be considered an implicit argument to any routines whose behaviour is affected.

The accuracy of the formulation is not essential for Lewis' argument; *if the size of an output-only program cannot be objectively determined, the addition of input, state, and interactivity will not modify the conclusion.*

Computing upper bounds for $K(x)$ is not difficult: compress the string x with some algorithm,[13] implement the corresponding decompressor in the chosen language, concatenate the decompressor to the compressed string, and measure the length of the resulting string, concretely, the size of a self-extracting archive in the given language.

Rephrasing the incomputability of K for this case shows no algorithm for finding the shortest program with a desired behaviour exists. Furthermore, a tight upper bound to K can be computed [377] but it is not feasible.

According to Vardi [403]

> Ultimately, software development is an engineering activity, not a mathematical activity. Engineering design and analysis techniques do not provide mathematical guarantee, they provide confidence. We do not need to solve the Halting Problem, we just need to be able to reason successfully about termination of real-life programs. It is time to give up our "unsolvability phobia". It is time to solve the unsolvable.[14]

4.5 Program behaviour

Modern computing systems allow many users to share the same facilities, memory, processors, databases, or software, such as compilers and rou-

[13] The Lempel-Ziv compression methods are among the most popular lossless algorithms [427]. Modern lossless compressors use probabilistic models, such as prediction by partial matching. The Burrows-Wheeler transform is an example; see [58].
[14] Challenge for the reader: Do you agree?

tines. This raises questions about protecting various objects from damage and misappropriation, leading to multiple problems regarding program behaviour.

For example, questions like "Will the instruction `Inst` be ever executed?" or "Is a high-security value stored in low-security memory?" are all about program behaviour; these are semantic properties. In contrast, questions like "Does this program contain a for-loop?" or "Is this program type-correct?" are not behaviours because there might be equivalent programs that have different answers for them. Indeed, one can write a program with a for-loop using while-loops without changing the behaviour, or you can make a correct version that doesn't respect specific imposed rules. The above questions refer to particular implementations. They are not about behaviour; they are syntactic properties.

The solutions proposed to answer questions about program behaviour are diverse, involving a mixture of hardware and software. With such diversity, it is fruitful to abstract some essential features of the systems and develop a formal model of protection.

An interesting example is the problem faced by system administrators when assessing the impact of an access control policy on their system's security. A formal "access control" model was presented in [215].

A state in the system is denoted by a set of objects, some of which are subjects, and a protection matrix giving the current access rights between pairs of subjects and objects. The Safety Problem can be formulated as follows:

> Given a set of policy rules, generic access right X and an initial matrix, is it possible to reach a state where X is granted to any subject?

The Safety Problem is undecidable because of a more general undecidability result – Rice Theorem[15] – which says that deciding every non-trivial behavioural property of a program (Turing machine) is undecidable.

Programs will be coded by strings called descriptions, see Section 2.8.

Rice Theorem.[16] *Let P be a language consisting of program descriptions fulfilling the following two conditions:*

[15]The theorem is named after H. G. Rice, who proved it in his doctoral dissertation of 1951 at Syracuse University [339].

[16]For a proof see Section 5.7.

(1) P is non-trivial, that is, it contains some, but not all, program descriptions,

(2) P is a property of the behaviour of the program (i.e. a property of the language of outputs it accepts/produces), as opposed to a particular syntactic feature of the implementation.

Then, P is undecidable.

An informal formulation is:

> Any decidable property of the behaviour of computer programs is trivial, i.e. it is always true or false.

Each of the following behaviour properties of programs is undecidable by Rice Theorem because they verify the conditions (1) and (2) above:

(1) the program accepts no input,
(2) the program accepts every input,
(3) the language of accepted inputs is infinite,
(4) the language of accepted inputs is finite,
(5) the program accepts some input.

Consider now an input w and the property P_w of programs defined as follows:

$$\text{the program } M \text{ stops on the input } w. \tag{4.1}$$

First, the set of programs stopping on the input w is undecidable by the Halting Theorem. Indeed, assume by the absurdity that the property P_w is decidable. Then we can use it to decide the Halting Problem: to decide, for an arbitrary program M and input w, whether M stops on the input w, check whether P_w is true or not. Hence, for every w the set of programs M stoping on w is undecidable. However, the undecidability does not follow from Rice Theorem: it is non-trivial but does not satisfy condition (2). Indeed, consider the program M_1 which stops on every input and the program M_2 which stops only on w: both programs have the property P_w, but P_w is not a behavioural property of programs as the set of outputs of M_1 is the set of all inputs while the set of outputs of M_2 consists of only $\{w\}$.

The argument presented in the above paragraph is called "reducibility": we have reduced the decidability of the problem (4.1) to that of the Halting Problem; because of the Halting Theorem, the problem (4.1) was proven undecidable.

Finally, the set of programs not stopping on the input w is undecidable by the Halting Theorem but not Rice Theorem.

4.6 Network insecurity

Internet insecurity has been spreading at Internet speed for more than 35 years. The Morris worm [409] is one of the oldest computer worms distributed via the Internet and the first to get mainstream media attention. R. Morris, a tenured professor at MIT since 2006, developed and released the worm on 2 November 1988 by exploiting several vulnerabilities to gain entry to targeted systems, which include a:

(1) hole in the debug mode of the Unix Sendmail program,
(2) buffer overflow or overrun hole in the fingerd network service,
(3) transitive trust enabled by people setting up network logins with no password requirements via remote execution (rexec) with Remote Shell (rsh).

Unfortunately, Morris's coding mistake, which instructed the worm to replicate itself despite a computer's reported infection status, transformed the code from an intentionally harmless computing exercise into a viral denial of service attack. The result was a felony conviction in the USA under the 1986 Computer Fraud and Abuse Act.

More attacks have followed: password sniffing (1994), IP spoofing (1995), denial of service attacks (1996), email-borne viruses (1999), distributed denial of service (2000) and many others with no end in sight. After many years of concerted effort, computer insecurity has only gotten worse.

The keywords in these scenarios are:

(1) *threats* are possible attacks,
(2) *vulnerabilities* are weaknesses,
(3) *assets* are information and resources that need protection,
(4) *risk* of threats and vulnerabilities.

The "wisdom" in this field is

Absolute security is impossible.

Are there any "objective" results justifying this empirical "wisdom"?

A step towards answering this question is to ask more specific questions such as the following ones [268]:

> Can a browser comprehensively block "unsafe" JavaScript for some reasonable security model, and what computational power would be required to do so?
>
> Does JSON promote safer Web app development?

Is information insecurity acceptable? The answer is negative, but this answer has no value unless specific measures accompany it. The ISM Document of the New Zealand Government Communications Security Bureau[17] is an example.

4.7 Infinite divisibility

In the 5th century BCE, the Ancient Greek pre-Socratic philosopher Democritus and his teacher Leucippus theorised matter's divisibility beyond what can be perceived by the senses until ultimately ending at an indivisible atom. Is time considered to be finitely or infinitely divisible?

According to The Standard Model,[18] the Planck length, which is equal to $1.616229(38) \times 10^{-35}$ metres, is the distance at which the Standard Model is expected to "break down". This is the smallest physical scale at which meaningful physical statements can be currently made. The time analogue is $5.39116(13) \times 10^{44}$ seconds, the Planck time. A new theory of quantum gravity, unifying quantum mechanics and general relativity, could predict the physical behaviour of space-time at smaller distances.

Assuming that space is infinitely divisible, Zeno argued [266, p. 45], that *there is no motion.* Indeed, if there is motion, then

> it is possible in a finite time to traverse an infinite number of positions, making an infinite number of contacts one by one; but this is impossible ...

For A. N. Whitehead, "an infinite number of acts of becoming may take place in a finite time if each subsequent act is smaller in a convergent series" [346, pp. 182–183]:

[17] `www.nzism.gcsb.govt.nz/ism-document`, from September 2022.
[18] `en.wikipedia.org/wiki/Standard_Model`.

The argument, so far as it is valid, elicits a contradiction from the two premises: (i) that in a becoming something (*res vera*) becomes, and (ii) that every act of becoming is divisible into earlier and later sections which are themselves acts of becoming. Consider, for example, an act of becoming during one second. The act is divisible into two acts, one during the earlier half of the second, the other during the later half of the second. Thus that which becomes during the whole second presupposes that which becomes during the first half-second. Analogously, that which becomes during the first half-second presupposes that which becomes during the first quarter-second, and so on indefinitely. Thus if we consider the process of becoming up to the beginning of the second in question, and ask what then becomes, no answer can be given. For, whatever creature we indicate presupposes an earlier creature which became after the beginning of the second and antecedently to the indicated creature. Therefore there is nothing which becomes, so as to effect a transition into the second in question

However, the experience shows *there is motion*![19]

Fig. 4.2: S. Wolfram[20]

[19]Zeno's argument is an early, if not the first, example of "proof by contradiction", also called *reductio ad absurdum*: to prove the truth or the validity of a statement, assume its falsity and deduce a contradiction. Proof by contradiction is an indirect and not constructive type of proof.

[20]Source: In Wikipedia. `en.wikipedia.org/wiki/File:Stephen_Wolfram_PR_(cropped).jpg`.

There have been "resolutions" of Zeno's paradox,[21] including Russell's [349]. Other resolutions use the assumption of infinite divisibility of both space and time; see [207]. The simplicity and universality of Zeno's arguments on the motion will continue to fascinate and inspire.

But what if the space is not infinitely divisible? If this were the case, then computability theory, based on discrete components, time, space, memory capacity, program length, etc. would apply directly to physics. Various studies on "digital physics" try to develop such a theory; for example, Wolfram's Physics Project [412, 411].

4.8 Is physics immune to incomputability?

Wigner's famous article [408] starts with the following Russell's citation:

> Mathematics, rightly viewed, possesses not only truth, but supreme beauty cold and austere, like that of sculpture, without appeal to any part of our weaker nature, without the gorgeous trappings of painting or music, yet sublimely pure, and capable of a stern perfection such as only the greatest art can show. The true spirit of delight, the exaltation, the sense of being more than Man, which is the touchstone of the highest excellence, is to be found in mathematics as surely as in poetry.

Mathematics plays a sovereign role in physics, and chaotic behaviour exists in many physical systems (see also the discussion about chaos in Section 2), so it is natural to ask whether *incomputable physical systems exist.*

Incomputability in physics is not a new topic. A selection of articles includes [414, 326, 244, 136, 137, 277, 142, 278, 142]; Svozil [385] is the first monograph dedicated to this subject, still an essential reference.

What does it mean to say that a physical system is computable? A physical system is determined by a finite set of initial conditions and laws represented by equations. The system is computable if we can solve the equations in all instances, using a "finite body of mathematical knowledge". In this case, if the laws of the system can be coded into a computer program, then given the system's initial conditions, one can compute the system's evolution.

[21]The origin of paradoxes is unclear. Bible's Epistle to Titus 1:12, states that "one of Cretes own prophets has said it: 'Cretans are always liars, evil brutes, lazy gluttons'. He has surely told the truth."

This scenario is tacitly taken for granted. But is this reasonable? The answer is negative, and a straightforward example is when the initial conditions or equations include irrational numbers, like π or Ω: no program can deal with such numbers because of their *infinity.*

In the immensely popular book *One Two Three ... Infinity* [188], G. Gamow proposed a scenario in which you live on the surface of a two-dimensional balloon and wish to detect its three-dimensional curved character using only two-dimensional observations. This is a non-trivial challenge. In this context Cooper [137] asked two important questions:

> Is there a causal dimension that is not algorithmic? Does it *matter* if there is?

Gamow proved that despite it being tricky to live in a two-dimensional space, one still could find evidence of the third dimension.[22] Furthermore, Cooper argued that *it did matter precisely because we could find that evidence.*

The German computer pioneer K. Zuse proposed the hypothesis that the universe is a gigantic universal cellular automaton, or equivalently, a universal Turing machine, in his 1969 influential book on automata theory [428]. This view is known today as the Simulation Hypothesis,[23] which is related to digital philosophy and digital physics; see [183, 184, 425, 424].

Would such a universe be computable? The answer seems affirmative. A universe given by a universal Turing machine, a finite object, should be computable. Such a universe appears computable "per definition", but only in a limited sense, on a step-by-step basis. "Global" questions such as those discussed in the Halting Problem, Specker Theorem and Rice Theorem point to incomputability. This difference was discussed in [384][24] and more recently in Longo [277].

The tests of incomputability discussed in Section 5.9 complement the theoretical arguments and suggest a negative answer to the question in this section title.

---

[22]Challenge for the reader: Find the evidence.

[23]This hypothesis was discussed recently in [122]. The Western University hosts the Centre for Digital Philosophy, `tinyurl.com/yckymtd8`.

[24]As noted by K. Svozil in an email to C. Calude on 8 February 2023.

4.9 The power and weakness of randomness

Slot machines are probably the most lucrative games on the casino floors for the house, players, and cheaters. These machines have always been a game of cat and mouse between cheaters and the house since they were invented, and the game runs at full speed. Regulatory bodies oversee the gambling industry to ensure its proper and equitable functioning. Gaming engineers create machines that can be monitored and audited for quality gameplay and compliance. However, what are the potential consequences if an engineer manipulates the codes for personal gain? This is an improbable but not impossible scenario. Indeed, R. D. Harris,[25] an engineer from The Nevada Gaming Commission, did precisely that. Over several years, he exploited the machines by accessing their source codes. The scheme only came to light when his accomplice won a significant sum of $100,000 while playing keno at a casino in 1995.[26]

Fast forward to the early days of June 2014, the Lumiere Place Casino in St. Louis accountants detected some irregular activity with a few of their slot machines over a few days. The software approved by the government to operate such machines is designed to give the house a fixed mathematical advantage, ensuring the casino's earnings are predictable in the long run, typically around 7.129 cents for every dollar played. However, on 2 and 3 June, several machines paid out much more than they had received without awarding any significant jackpots. This anomaly, a negative hold in the industry, suggested that someone was cheating. And indeed, the casino's security identified the culprit: a man in his thirties with black hair who wore a zip-up Polo shirt and carried a square brown purse. Unlike other slot machine cheaters, he didn't tamper with the machines; instead, he played games like Star Drifter or Pelican Pete, pressing the buttons while discreetly holding his iPhone close to the screen. After a few minutes of play, he would leave the machine and return later for a second try. On the second attempt, he would often win, turning a small investment of $20 to $60 into as much as $1,300 before cashing out and moving to another machine to start the cycle anew. Over the two days, he had racked up a total of over $21,000 in winnings. The only unusual aspect of his behaviour was how he would hover his finger above the Spin button for extended periods before quickly pressing it, unlike regular slots players who tend to spin continuously without pause.

[25] gaming.nv.gov/index.aspx?page=190.

[26] www.casino.org/blog/keno-cheater-got-busted/.

A statewide alert was issued, and it was soon revealed that several other casinos had fallen victim to the same scheme, albeit often by different individuals. In each case, the culprit had held a cell phone near an Aristocrat Mark VI model slot machine just before a winning streak. Missouri authorities identified the individual behind the Lumiere Place scam as M. Bliev, a 37-year-old Russian. Bliev had flown back to Moscow, but the St. Petersburg-based organization[27] he was affiliated with, sent him back to the USA to work with another cheating team [251] – a big mistake!

The entertaining book [357] exposes the hackers' toolkits and answers vital questions: Why is the Internet so vulnerable? What can we do in response? How did they do?

As of early 2011, European casinos began reporting instances in which slot machines produced by the Austrian firm Novomatic paid out vast sums. Novomatic's engineers found no indications of tampering, hence they hypothesised that the cheaters had algorithms to predict the behaviour of the slot machines.

Slot machine outcomes are controlled by Pseudo-Random Number Generators (PRNGs)[28] that are vetted by Government regulators before casinos can deploy them. The suspicion was that the "quality" of the random numbers produced was not good enough. So, what is a random number?

Random numbers have existed for over 4,000 years, but never have they been in such demand as in our time. The oldest known dice were discovered in a 24th century BC tomb in the Middle East.

Fig. 4.3: Kleroterion by Marsyas – Own work, CC BY-SA 2.5[29]

[27]Which earns millions by cracking gaming industry algorithms.

[28]A PRNG is a program which produces strings of bits which exhibit *some* statistical properties of randomness.

[29]Source: In Wikipedia. `upload.wikimedia.org/wikipedia/commons/a/a5/AGMA_Kleroterion_1.jpg`.

Kleroterion was a randomisation device used by the Athenian city to democratically select citizens for most state offices and court juries. Bronze identification tickets were inserted to indicate eligible jurors divided into tribes. Using Kleroterion – which was put in front of each court – a whole row would be accepted or rejected for jury service.[30]

In China, as early as 1100 BC, fortune tellers used heated turtle shells until they cracked randomly and interpreted the cracks to predict the future.

In 1949 Reichenbach [336] was the first to claim that humans cannot produce a random series of responses. Later studies concluded that humans are not good at generating random numbers [406, 180]. One cause seems to be that humans alternate too much between choices. For example, experiments showed that in 60% of cases, humans place 1 after a 0, or vice-versa, when in a proper good random sequence, they alternate in around 50% of cases (which is also not good if the alternation is too "rigid"; see, for example, the Champernowne sequence defined in (5.5)).

Statistician F. Galton wrote in 1980 [187]:

> As an instrument for selecting at random, I have found nothing superior to dice.

J. von Neumann developed the first PRNG around 1946 using the following simple algorithm:

> start with an initial random seed value, square it, and slice out the middle digits.

He was also the first to recognise the pitfalls of PRNGs:

> Anyone who considers arithmetical methods of producing random digits is, of course, in a state of sin.

In 1951, Ferranti Mark 1 was the first computer with a built-in random number instruction – designed by A. Turing – that could generate 20 random bits at a time using electrical noise.[31]

[30] This method, called *sortition*, has been used in Belgium parliaments since 2017: Belgian Sortition Models – Institutionalizing Deliberative Democracy, `congress.crowd.law/case-belgian-sortition-models.html`.

[31] C. Strachey used it to program a random love generator. D. Link resurrected Stracheyś program in 2009, `http://www.alpha60.de/art/love_letters/` and here is a letter: DUCK SWEETHEART, MY CURIOUS EAGERNESS FONDLY HUNGERS FOR YOUR DEVOTION. YOU ARE MY BURNING PASSION. MY AMBITION PRIZES

RAND Corporation generated numbers using a random pulse generator: its book *A Million Random Digits with 100,000 Normal Deviates* was published in 1955 [393] and it is still available in paperback on Amazon.

RDRAND ("read random"), known as Intel Secure Key Technology, is an instruction returning random numbers from an Intel on-chip hardware random number generator which has been seeded by an on-chip entropy source.[32]

Pseudo-random numbers are used hundreds of billions of time daily to encrypt electronic network data. They are used in various applications in science, economy and security, ranging from cryptography to numerical simulations. Their pitfalls have been discovered for more than 30 years. One of, if not the first example, was reported in July 1992 [179]. The abstract summarises it all:

> The Wolff algorithm is now accepted as the best cluster-flipping Monte Carlo algorithm for beating "critical slowing down." We show how this method can yield *incorrect* answers due to subtle correlations in "high quality" random number generators.

Other drawbacks have been discovered in the Internet era: in 2012, a flaw in the encryption system RSA used worldwide for online shopping, banking, and email [290] was traced to the numbers a PRNG has produced [269]. The use of RDRAND in a Monte Carlo simulator was evaluated with a focus on performance and reproducibility: the conclusion was that using RDRAND as opposed to Mersenne Twister doesn't provide different results, but worse [347]. In 2020 Vrije Universiteit Amsterdam researchers released a report on a side-channel attack named CrossTalk (CVE-2020-0543) [334] affecting RDRAND in Intel processors. This vulnerability affects situations where untrusted code is run with trusted code on the same processor, as in a shared hosting environment. T. Murphy's *Pseudorandom Hall of Shame* includes many examples of PRNG's failures.[33]

What is the reason that pseudo-randomness is affected by many pitfalls? Sequences of pseudo-random digits follow subtle patterns that can be observed over long periods and many instances. There are correlations and patterns "generated" by the algorithm used to produce them. "Good" randomness is the critical resource for all these applications, but high-quality

---

YOUR UNSATISFIED FELLOW FEELING. MY ARDOUR PASSIONATELY LIKES YOUR LONGING. MY LOVING PASSION TREASURES YOUR HEART.

[32] https://en.wikipedia.org/wiki/RDRAND.

[33] Communicated in an email to C. Calude.

randomness is hard to produce and certify. What is high-quality randomness? The quality of randomness can be defined and evaluated, as we will discuss in Section 4.12, but first, we note that

there is no true/perfect/total randomness.

What does it mean, and how to justify it? True/perfect randomness means a total lack of correlations. Some religious believers hold the view that there is no true randomness in the world, as everything is part of a divine plan or guided by a higher power. A different answer will be given in the next section.

4.10 Relational databases, Ramsey theory and the impossibility of complete disorder

E. F. Codd invented the notion of a relational database at IBM in 1970 [128].

A relational database D is a collection of data and relationships between them. The relational database is organised into one or more tables of columns and rows, with a unique key identifying each row. Tables contain information about the objects to be represented in the database. Each column in a table holds a particular type of data, and a field stores the actual value of an attribute. The rows in the table represent a collection of related values of one object or entity. Data can be accessed in many ways without reorganising the database tables.

Formally, a correlation of variables in D with sets of n elements is a set B of size b whose sets form the correlation. In other words, when a correlation function – defined according to some given criteria (proximity or, conversely, apartness of some observable values or whatever) – selects a set of n elements, forming a set of cardinality b, then they become correlated. Thus, the selection process may be viewed as colouring the chosen set of b elements with the same colour – out of $c > 1$ possible ones. The binary case $c = 2$ is the simplest.

The criterion of selection – the correlation function – has no relevance here; it is arbitrary. It only matters that, for some reason or even no reason, all sets of n elements of a set with b elements have the same colour, that is, turn out to be correlated.

We have written "for some reason or even no reason". Does this suggest that some correlations have no "meaning"? In the simplest case, the question is: Are there binary strings with *no* patterns/correlations? *Ramsey theory*[34] *answers the question in the negative.*

Fig. 4.4: F. P. Ramsey[35]

Significantly, Graham and Spencer subtitled their *Scientific American* presentation of Ramsey theory [204] with the following sentence:

> Complete disorder is an impossibility. Every large set of numbers, points or objects necessarily contains a highly regular pattern.[36]

In what follows, we will use two Ramsey-type results. In both cases the *Ramsey-type of correlations appear only because of the size of the data,* with no relation to meaning.

We start with an example. Let $s_1 \ldots s_n$ be a binary string. A monochromatic arithmetic progression of length k is a substring $s_i s_{i+t} s_{i+2t} \ldots s_{i+(k-1)t}$, for $1 \leq i$ and $i + (k-1)t \leq n$ with all characters equal for some $t > 0$. The string 01100110 contains no arithmetic progression of length 3 because the positions 1, 4, 5, 8 (for 0) and 2, 3, 6, 7 (for 1) do not contain an arithmetic progression of length 3. However, both strings 011001100 and 011001101 do: 1, 5, 9 for 0 and 3, 6, 9 for 1 and this change was obtained by adding just one bit. In fact, *all 512 binary strings of length 9 have a monochromatic arithmetic progression of length 3.*[37]

[34] Named after the British mathematician and philosopher F. P. Ramsey.
[35] Source: In Wikipedia. en.wikipedia.org/wiki/File:Frank_Plumpton_Ramsey.JPG.
[36] A measure-theoretical argument is given in [61, p. 171].
[37] Challenge for the reader: Prove it.

Fig. 4.5: B. L. Van der Waerden[38]

The theorem below states that *all* sufficiently long strings of digits/colours, taken from a finite set, have "long enough" arithmetic progressions of the same digit/colour, i.e. a sequence of equidistributed positions of the same digit/colour (monochromatic) *of a pre-given length* [203]. The importance of the theorem lies in the fact that *all* strings display one of the simplest types of correlation: a sequence of equidistributed positions of the same value.

Van der Waerden Theorem. *For any positive integers k and c there is a positive integer γ such that every string, made out of c digits (or colours), of length more than γ contains an arithmetic progression with k occurrences of the same digit or colour, i.e. a monochromatic arithmetic progression of length k.*

We illustrate the phenomenon with the classic "party problem".[39] Suppose a party has six people. Consider any two of them. They might be meeting for the first time, in which case we call them *mutual strangers*, or they might have met before, in which case we call them *mutual acquaintances.*[40] The problem asks whether *in every party of six people either: a)*

[38]Source: In Wikipedia. `en.wikipedia.org/wiki/File:ETH-BIB-Waerden,_Bartel_Leendert_van_der_(1903-1996)-Portrait-Portr_12193.tif`.

[39]Appeared in the Putnam Mathematical Competition in 1953 and the problem section of the *American Mathematical Monthly* in 1958 (Problem E 1321).

[40]This is a Ramsey type problem: the aim is to determine how large the party needs to be to guarantee similar pairwise acquaintanceship in (at least) one group of three people.

at least three of them are (pairwise) mutual strangers or b) at least three of them are (pairwise) mutual acquaintances. The problem is equivalent to the statement that in the complete graph in Fig. 4.6 – which has six vertices and every pair of vertices is joined by an edge – every colouring of edges with red and blue necessarily contains either a red triangle or a blue triangle.

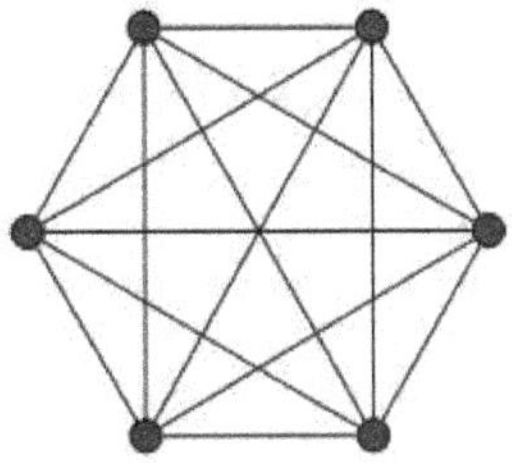

Fig. 4.6: Party graph

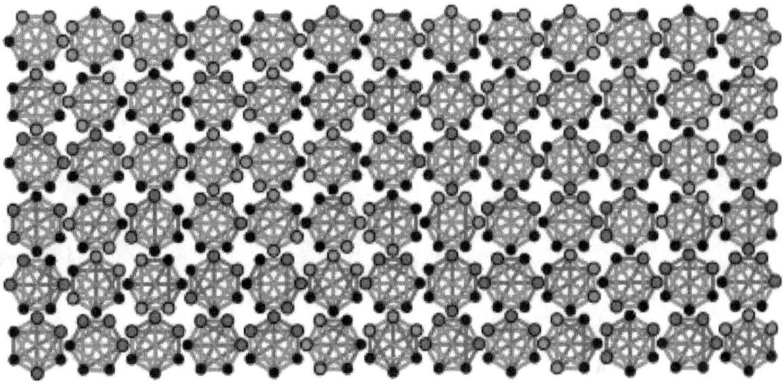

Fig. 4.7: Half of the friends-strangers graphs with 6 nodes[41]

The answer is *affirmative*. To prove this, we must choose any vertex in Fig. 4.6, and note that five edges leave it, each coloured red or blue. Then the *Pigeonhole Principle* says that at least three of them must have the same colour: indeed, if there are less than three of one colour, say red, then there are at least three that are blue. Hence a monochromatic triangle will

[41]Source: In Wikipedia. `en.wikipedia.org/wiki/File:Friends_strangers_graph.gif`.

appear. In each graph in Fig. 4.7, the red/blue nodes show a sample triplet of mutual friends/strangers.[42]

We *do not know* whether at the party there is a group of three people who are pairwise mutual strangers or a group of three people who are pairwise mutual acquaintances, but we *know* that a group of three people with the same type of pairwise acquaintanceship always exists. This result is true for any party with more than six people, but not for less.[43]

The Van der Waerden number $W(k, c)$ is the smallest γ such that every string, made out of c digits or colours, of length $\gamma + 1$ contains a monochromatic arithmetic progression of length k. How big is $W(k, c)$? One has $W(3, 2)=9$ for binary strings. In [201], it was proven that

$$W(k,2) < 2^{2^{2^{2^{2^{k+9}}}}},$$

and conjectured to be much smaller in [202]: $W(k, 2) < 2^{k^2}$.

In the article [204], Graham and Spencer cite the great mathematician P. Erdös:

> Suppose aliens invade the earth and threaten to obliterate it in a year's time unless human beings can find the Ramsey number for red five and blue five. We could marshal the world's best minds and fastest computers, and within a year we could probably calculate the value. If the aliens demanded the Ramsey number for red six and blue six, however, we would have no choice but to launch a preemptive attack.

This means that if the aliens are looking for $W(5, 5)$, humanity should divert all of its computing resources to figure out the answer. However, if they demand the value for $W(6, 6)$, humans should prepare for war.

In March 2023, four Brasilian mathematicians, Campos, Griffiths, Sahasrabudhe and Morris posted a proof for the improved inequality [112]:

$$W(k,2) < 3.993^k < 4^k.$$

The 56-page proof will take weeks for every detail to be fully verified by the combinatorics community. But experts are optimistic.[44]

For the next result, we need a bit of notation. If A is a finite set, then $[A]^n$ is the set of all subsets of A containing exactly n elements, n-sets on A. To guide intuition, the number c may also be viewed as the

[42] Challenge for the reader: Are these all graphs?

[43] Challenge for the reader: Why?

[44] According to Wikipedia page "Ramsey's theorem", https://tinyurl.com/5f5ft5p6, the proof was not validated as of 30 June 2023.

number of "colours" partitioning a set. For example, if $A = \{x, y, z, u\}$ and $n = 2$, then $[A]^2 = \{\{x, y\}, \{x, z\}, \{x, u\}, \{y, z\}, \{y, u\}, \{z, u\}\}$; an example of partition of $[A]^2$ with $c = 3$ colours, red, blue and yellow, is $\{\{\{x, y\}, \{x, z\}, \{z, u\}\}, \{\{y, z\}, \{y, u\}\}, \{\{x, u\}\}\}$ (if $\{x, y\}, \{x, z\}, \{z, u\}$ are coloured with red and $\{y, z\}, \{y, u\}$ are coloured with blue, then $\{x, u\}$ is coloured with yellow). Any of the six permutations of red, blue and yellow produce valid colouring.

> **Ramsey Theorem.** *For all positive integers b, n, c, there is a positive integer γ such that for every finite set A containing more than γ elements and for every partition $P\colon [A]^n \to \{1, 2, \ldots, c\}$, there exists a subset B of A containing b elements whose n-sets are monochromatic, i.e. $P(x)$ has the same value (colour) for every x in $[B]^n$.*

The Ramsey Theorem shows that, given any correlation function and any b, n and c, there always exists a large enough number γ such that *any* set A of size greater than γ contains a set B of size b whose subsets of n elements are all correlated – that is, monochromatic. In other words: Do we want a size b set of values correlated by sets of n elements out of c possibilities for whatever b, n, and c *we* choose? Ramsey Theorem gives us a large enough number γ such that in *any* set with more elements than γ, if we choose in *any* way a partition of n-sets into c classes, we are guaranteed to find a correlation – called *Ramsey correlation* – of size b and arity n. We do not know *a priori* what the colour of the monochromatic set will be, in the same way, the data miner does not know in advance which correlation will pop out from the data. However, we can algorithmically find all the monochromatic elements and their colourings in every particular instance.

The analysis above, as well as the one in the following section, are independent of any possible (or not) law-like origin (determination) of the processes modelled in the database, as *they are only based on the size of the database.*

4.11 Spurious correlations in big data: "Too much information tends to behave like very little information"

Sastry Pantula was named the head of the Division of Mathematical Sciences (DMS) at the US National Science Foundation (NSF) in May 2010, and he took over the leadership of NSF/DMS in September 2010. The same year, Professor Pantula proposed changing the DMS's name to the Division of Mathematical and Statistical Sciences (DMSS). His letter motivating the proposal includes the following paragraph (cf. [22, p. 933]):

> Big data provide big opportunities for mathematical and statistical sciences. It is an exciting time for our Division. In his FY12 budget roll-out speech NSF Director Dr. Subra Suresh referred to the "era of data and observation". The NSF 20112016 Strategic Plan notes that "The revolution in information and communication technologies is another major factor influencing the conduct of 21st century research. New cyber tools for collecting, analyzing, communicating, and storing information are transforming the conduct of research and learning. One aspect of the information technology revolution is the 'DATA DELUGE', shorthand for the emergence of massive amounts of data and the changing capacity of scientists and engineers to maintain and analyze it." Extracting useful knowledge from the deluge of data is critical to the scientific successes of the future. Data-intensive research will drive many of the major scientific breakthroughs in the coming decades. There is a long-term need for research and workforce development in computational and data-enabled sciences. Statistics is broadly recognized as a data-centric discipline, thus having it in the Division's name as proposed would be advantageous whenever "Big Data" and data-sciences investments are discussed internally and externally.

As noted by Andrews [22, p. 934], "There is remarkable restraint in the paragraph by Pantula ... in contrast with the widespread giddy enthusiasm for very large databases and data analytics." The article by C. Anderson in *Wired Magazine* [21], from where we quote the following paragraphs, is the best illustration of this excitement:

> "All models are wrong, but some are useful." So proclaimed statistician George Box thirty years ago. ... Peter Norvig, Google's research director, offered an update to George Box's maxim: "All models are wrong, and increasingly you can succeed without them."[45]
>
> This is a world where massive amounts of data and applied mathematics replace every other tool that might be brought to

[45] Anderson attributed the last sentence to Google's research director Peter Norvig, who denied it [308]: "That's a silly statement, I didn't say it, and I disagree with it."

> bear ... With enough data, the numbers speak for themselves. ...
>
> The scientific method is built around testable hypotheses. These models, for the most part, are systems visualized in the minds of scientists. The models are then tested, and experiments confirm or falsify theoretical models of how the world works. This is the way science has worked for hundreds of years. Scientists are trained to recognize that correlation is not causation, that no conclusions should be drawn simply on the basis of correlation between X and Y (it could just be a coincidence). Instead, you must understand the underlying mechanisms that connect the two. Once you have a model, you can connect the data sets with confidence. Data without a model is just noise. But faced with massive data, this approach to science – hypothesize, model, test – is becoming obsolete. Consider physics: Newtonian models were crude approximations of the truth (wrong at the atomic level, but still useful). A hundred years ago, statistically based quantum mechanics offered a better picture but quantum mechanics is yet another model, and as such it, too, is flawed, no doubt a caricature of a more complex underlying reality. The reason physics has drifted into theoretical speculation about n-dimensional grand unified models over the past few decades (the "beautiful story" phase of a discipline starved of data) is that we don't know how to run the experiments that would falsify the hypotheses – the energies are too high, the accelerators too expensive, and so on. Now biology is heading in the same direction. ... In short, the more we learn about biology, the further we find ourselves from a model that can explain it. There is now a better way. Petabytes allow us to say: "Correlation is enough." We can stop looking for models. We can analyze the data without hypotheses about what it might show. We can throw the numbers into the biggest computing clusters the world has ever seen and let statistical algorithms find patterns where science cannot.

The article ends with the call:

> Learning to use a "computer" of this scale may be challenging. But the opportunity is great: The new availability of huge amounts of data, along with the statistical tools to crunch these numbers, offers a whole new way of understanding the world. Correlation supersedes causation, and science can advance even without coherent

models, unified theories, or really any mechanistic explanation at all.

There's no reason to cling to our old ways. It's time to ask: What can science learn from Google?

The idea behind this new radical "philosophy" is that sufficiently powerful algorithms can now explore massive databases and can find therein correlations and regularities. Independently of any analysis of the "meaning" or "content" of such correlations – which are notions challenging to define – the machine provides rules for prediction and action. The strength and generality of this method rely on the immense size of the database: the larger the data, the more powerful and effective is the method grounded on computationally-discovered correlations. Hence, there is no need to theorise, understand, or criticise the sense of the discovered correlations: "No semantic or causal analysis is required", "Correlation is enough .... We can throw the numbers into the biggest computing clusters the world has ever seen and let statistical algorithms find patterns where science cannot".

The picture proposed by Anderson has generated immense excitement, and, to some extent, in the last decade, it has become, at least in part, reality. Big data is a major opportunity for science, and data analytics is a remarkable new field of investigation in computer science. AI and big data obtained astonishing results using processes and technologies to combine and analyse massive datasets to identify patterns and develop actionable insights. Faster data-driven decisions increase knowledge, efficiency, revenue and profits.

Correlation is at the core of this paradigm shift. But what is correlation? Co-related phenomena connect covariantly, that is, they vary while preserving the proximity of values according to a pre-given measure. "Co-relation" is essentially "co-incidence", that is, things that occur together. Assume we have two correlated variables taking (numeric) values. Then, if one variable value increases, the other increases (or decreases) "in the same way". For example, "as the temperature goes up, ice cream sales also go up" or "as the temperature decreases, the speed at which molecules move decreases" or "as attendance at school drops, so does achievement" or "as the weather gets colder, air conditioning costs decrease".

Correlations can be useful because of their potential predictive power: *use or act on the value of one variable to predict or modify the value of the other.* The importance of correlations has been recognised for a long time. Early astronomers in the region from Korea and China to Morocco detected various regularities, from which they could compute future observations of astronomical bodies, from about the 10th century BC onwards. The observations and computational methods gradually improved over the millennia so that by about 1200, astronomers could make reliable predictions of solar eclipses [303]. In the 10th century AC, Chinese doctors discovered that mild smallpox infection could prevent more severe illnesses; they confirmed this correlation by successfully experimenting with smallpox inoculation on large numbers of individuals [306, pp. 114–174]. These early observations of non-obvious correlations and their practical relevance have been improved by using the scientific method: a) Clairaut's mathematical derivation of the orbit of Halley's Comet, and its delays, on the grounds of Newton's laws and equations, in the 18th century, b) the understanding of microbial infections and vaccines in the 19th century.

However, not all people were convinced by Anderson's viewpoint and predictions: The article [22] is an early critique. Pitfalls of exaggerating prediction value based on correlated observables have been discussed in the literature for many years. For example, the conclusion of Ferber's 1956 analysis [178] is:

> Clearly the coefficient of correlation is not a reliable measure for [the practical problem of selecting functions (hypotheses) for predictive purposes], nor does there appear to be at the present time any alternative single statistic adequate in this respect relating to the period of observation. Unsettling as it may seem, there does not appear to be any statistical substitute for *a priori* consideration of the adequacy of the basic hypothesis underlying a particular function.

Despite long-known facts pointing to the contrary, the core of the argumentation rests on the belief that *correlation supersedes causation*, a viewpoint largely discussed in the literature. For example, Poppelaar's blog from 5 April 2015, titled "Do numbers really speak for themselves with big data?" [321] concludes that

> With data, enough computing power and statistical algorithms patterns will be found. But are these patterns of any interest? Not many of them will be, as spurious patterns vastly outnumber the meaningful ones.[46] Anderson's recipe for analysis lacks the scien-

[46]See the book [404] based on a huge collection of spurious correlations: the old rule that "correlation does not equal causation" is illustrated through hilarious graphs.

> tific rigour required to find meaningful insights that can change our decision making for the better. Data will never speak for itself, we give numbers their meaning, the Volume, Variety or Velocity of data cannot change that.

Correlation does not imply causation: this fact has been established for a long time. To understand it, we will work with the following informal definitions:

(1) The variables X and Y are correlated when they have a "tendency" to change and move together, either in a positive or negative direction.
(2) The variable X causes Y when a change in X leads to a change in Y. In other terms, the change in Y would not have happened without the preceding change in X.

If X and Y seem to be "linked", it is possible but inconclusive that X caused Y: other possibilities are that Y caused X or that some other variable Z caused both X and Y. How can we distinguish between these cases?

But do we need to distinguish between correlation and causation? The following example [312] with consequences worldwide is a good illustration:

> A 2010 study [337] conducted by Harvard economists Carmen Reinhart and Kenneth Rogoff reported a correlation between a country's economic growth and its debt-to-GDP ratio. In countries where public debt is over 90 percent of GDP, economic growth is slower, they found. The study gathered a lot of attention. Google Scholar listed 1218 citations at the time of writing, and the statistic was used in US economic policy debates in 2011 and 2012 [354][47].
>
> The study isn't conclusive, in fact, it's far from it. As John Irons and Josh Bivens of the Economic Policy Institute noted, the effect may run the other way around, with slow growth leading to high debt. Even more worryingly, the research didn't hold up on replication [218]. But by the time those facts became clear,

[47] European policymakers largely referred to that paper till 2013. For example, O. Rehn, EU Commissioner for Economic Affairs (2009–13) referred to the Reinhart-Rogoff correlation as a key guideline for his past and present economic views ([364], address to ILO, April 9, 2013) and G. Osborne, British Chancellor of the Exchequer (since 2010), claimed in April 2013: "As Rogoff and Reinhart demonstrate convincingly, all financial crises ultimately have their origins in one thing [the public debt]". [286].

the original study had already attracted widespread attention and affected policy decisions.

This example is not only about the type of correlations but also the *quality and "honesty" of collected data.* The analysis [206] shows the bias of Reinhart and Rogoff's collection of data in several countries for 218 years, hence its influence on their analysis.

Many hilarious real examples of high correlations can be found in `www.tylervigen.com/spurious-correlations`:[48]

(1) People who drowned after falling out of a fishing boat correlates with marriages rate in Kentucky, with a correlation $r = 0.952407$,
(2) US spending on science, space, and technology correlates with suicides by hanging, strangulation and suffocation, with a correlation $r = 0.99789126$,
(3) Total revenue generated by arcades correlates with computer science doctorates awarded in the US, with a correlation $r = 0.985065$.
(4) Math doctorates awarded in the US correlate with uranium stored at US nuclear power plants with a correlation $r = 0.952257$.

A critical analysis of spurious correlations was presented in [99]. The investigation uses three mathematical theories: a) ergodic theory to show that one cannot reliably predict by "analogy" with the past, even in deterministic systems, chaotic or not, b) Ramsey theory to prove that, given any arbitrary correlation on sets of data, there exists a large enough number (size) such that *every* data set larger than that size realises that type of correlation, c) algorithmic information theory to define a stand-alone definition of *spurious correlation* and to prove that the size of this set of correlations is far larger than the number of meaningful correlations – see Section 5.12.

Next, we use the Ramsey theory to illustrate spurious correlations presented in Section 5.12. Let D be a relational database.

> A correlation of variables in D is a set B of size b whose sets of n elements form the correlation – n values that are considered to be correlated.

In other words, when a correlation function – defined according to some given criteria (proximity, for example) – selects a set of n-sets, whose el-

[48]Checked on 21 August 2023.

ements form a set of cardinality b, they become correlated. The selection process may be viewed as colouring the chosen set of b elements with the same colour – out of c possible ones.[49]

As we mentioned before, *the selection criteria – used by the correlation function – have no relevance.* It only matters that all n-sets of a set with b elements have the same colour, that is, turn out to be correlated. Then, the Ramsey Theorem shows that, given any correlation function and every b, n and c, there always exists a large enough number γ such that *every* set A of size greater than γ contains a set B of size b whose subsets of n elements are all correlated – that is, monochromatic.

We do not know *a priori* what the colour of the monochromatic set will be. However, we can algorithmically find all the monochromatic elements and their colourings in every particular instance.

As a consequence, *every* huge database contains arbitrary correlations that appear only due to the database's size, not its nature. These correlations occur in the "randomly" generated databases of the required size, which implies that they are *spurious correlations.*

Anderson and his followers proclaimed the "end of science":[50] the reasons include that science is founded on models which are relatively arbitrary and often wrong; it is based on adductive and deductive praxis, the excessive critical thinking leaves room even for aesthetic judgements. The conclusion of the article [99]

> The scientific method can be enriched by computer mining in immense databases but not replaced by it.

it is still valid, and there is no indication that it can be contradicted soon.

Do we see now the "end of science"? AI with big data is a *disruptor* which dominates the good [41] and bad[51] news while science seems to be less innovative than a hundred years ago [313]:

> Recent decades have witnessed exponential growth in the volume of new scientific and technological knowledge, thereby creating conditions that should be ripe for major advances. Yet contrary to this view, studies suggest that progress is slowing in several major fields ... We find that papers and patents are increasingly less likely to

[49]The simplest case is $c = 2$.

[50]They are not the first to make this prediction; see, for example [229].

[51]Pause Giant AI Experiments: An Open Letter, `futureoflife.org/open-letter/pause-giant-ai-experiments/`.

break with the past in ways that push science and technology in new directions. This pattern holds universally across fields and is robust across multiple different citation- and text-based metrics.

Fig. 4.8: G. Longo[52]

A more positive opinion[53] was presented by the physicist R. Sheldon:

> When Einstein attended the Solvay conference in 1927, everyone there was a disruptor, an innovator. And everyone there did his science without government funding. Today, it would be hard to find a single scientist in a conference of 1000 who had not received government funding. And with the funding come the strings that hold back disruptors. There are undoubtedly many other factors contributing – the rise of the research university, the arrival of peer review, the publish or perish criteria, the sheer number of career scientists, the exponential rise in publications – but in my mind, it is the government control that is the dominating factor.

4.12 Quantum randomness

Randomness is powerful.[54]

[52]Source: G. Longo, email to C. Calude on 3 June 2023, `www.di.ens.fr/users/longo/Photos/photos-offic.html`.
[53]`evolutionnews.org/2023/01/is-the-end-of-science-near/`.
[54]Challenge for the reader: Why?

In Section 4.10, we showed that perfect/true/total randomness does not exist. This impossibility is not essential for many cases where random numbers are required: starting a tennis game with a coin toss is fine.[55] With a sample size of just 400 people chosen "randomly" from a population that exceeds 300 million, presidential polls conducted in the USA estimated the level of support for the last two candidates with an accuracy of within 5% while still having a 95% confidence level. Does the quality of randomness used in these polls matter? Not sure, but probably not.

(1) Randomness is crucial for secure encryption techniques that allow for confidential communication, even if an adversary has access to all coded messages.
(2) Random numbers can simplify enormous calculations while providing precise simulations for complex systems like the climate or the economy.

Algorithmic information theory [115, 61, 164] is a mathematical theory that studies the degrees of *algorithmic randomness* of individual objects, like strings and infinite sequences. While there is no true randomness, there is an increasing, unbounded scale of more and more algorithmically random objects. A handy tool for assessing the degrees of randomness is Martin-Löf tests of randomness [291, 61, 164]. Pseudo-randomness produced by PRNGs sits at the bottom because algorithms produce it, some better than others, but all predictable.[56] Martin-Löf randomness [291] is probably the most interesting form of algorithmic randomness because it combines a weak form of computability with strong "symptoms" of randomness.

Can algorithmic information theory help assess various forms of randomness produced in Nature? This theory is not useful for pseudo-randomness but for "incomputable" objects. So, *do such objects exist outside mathematics?* In other words:

> Can incomputable outputs be obtained from physical, chemical, or biological systems?[57]

There exist many devices that generate random numbers from physical processes, including:

---

[55]The detailed analysis in the article [161] concludes that the chance of a coin landing the same way it started is about 0.51.
[56]For example, the digits of π, PCG, Random123, xoroshilro128+, SHA-256.
[57]For randomness in biology; see [54, 98].

- macroscopic, e.g. coin, dice, roulette wheels, lottery machines,
- microscopic, e.g. thermal noise, photoelectric effect, quantum effects.

However, quantum mechanics is the most likely domain to look for "genuine" processes producing randomness. Why? Because quantum indeterminacy is the apparent incompleteness in describing a quantum physical system. For example, quantum mechanics claims that one cannot predict with certainty whether a radioactive atom will decay within a specified period, even with complete knowledge of the "laws of physics" and the atom's initial conditions. The best one can do is to calculate probabilities.

Indeterminacy in measurement had been established before quantum mechanics: experimental physicists found that measurement errors may lead to indeterminate outcomes. In the latter half of the 18th century, it was known that measurement errors could either be reduced by using better pieces of equipment or accounted for by statistical error models. However, indeterminacy in quantum mechanics has nothing to do with errors or disturbance.

The **QHypothesis** of irreducible randomness [421] stated by A. Zeilinger, one of the 2022 Nobel Prize winners in physics, states:

> ... for the individual event in quantum physics, not only do we not know the cause, there is no cause.

Fig. 4.9: A. Zeilinger[58]

[58]Source: In Wikipedia. `en.wikipedia.org/wiki/File:Godany_zeilinger2011_2452_blackboard.jpg`.

According to the **QHypothesis**, the essential feature of quantum randomness resides in its ontological *indetermination*, in contrast to the epistemic, subjective [217] uncertainty, which is reducible to ignorance of classical physical states and evolution, even in the case of chaotic behaviour.

There are several methods to generate quantum randomness, among them radioactive decay [253, 255], diodes [42], and quantum coin toss [385]. While the basic principle of quantum coin tosses as a source of "randomness" was already widely acknowledged, Svozil's paper presents a write-up of several discussions held at Vienna coffee houses, which involved active participation from A. Zeilinger with a strong belief in the irreducible indeterminism of such processes [421].

Various quantum random number generators (QRNGs) have been proposed and studied over the last decades: one, if not the first, was the quantum coin toss.

The interest in QRNGs has increased in the last decade because higher quality of randomness is required in many areas, from cryptography, statistics, and information science to medicine, physics, politics and religion,[59] and the many pitfalls of PRNGs, see Section 4.9, sometime catastrophic [290].

QRNGs are generally considered to be "better than PRNGs" because of a variation of the **QHypothesis** cited above [233], namely as a result of the

> "fundamental unpredictability of well-chosen and controlled quantum processes,"

a rather imprecise argument.

The mathematical results proving the impossibility of true/perfect randomness, see Section 4.10, have been ignored for many years in the quantum community. For example, the work reported in *Nature*[60] claims that quantum randomness is true randomness. More recently, an article published in July 2019 in *Wired* magazine[61] claims in the title: "Quantum Computers Could Be True Randomness Generators. Pure, verifiable randomness is essential to encryption yet hard to come by. Quantum computers could be

[59] For example, randomness is seen as an illusion or a limited human understanding of the greater order imposed by a deity. Free will and randomness will be discussed in Section 4.15.

[60] `doi:10.1038/news.2010.181`, 14 April 2010.

[61] `www.wired.com/story/quantum-computers-could-be-true-randomness-generators.`

the answer".

Quantum randomness, postulated by the **QHypothesis**, is generally reduced to the indeterminism of quantum measurements:

> Because the outcome is indeterministic, there is no way to predict it, hence it is random.

However, indeterminism does not imply randomness, and randomness does not imply indeterminism: they may co-exist. We illustrate this statement with pseudo-randomness generated by algorithms and the Omega number which is defined deterministically in equation (2.4).

> Is quantum randomness provably better than pseudo-randomness?

The answer seems to be affirmative: "We are sure!" Because, according to NIST, their *new quantum method generates really random numbers*,[62] (see [320, 47]) and the justification is:

> Quantum mechanics provides a superior source of randomness because measurements of some quantum particles (those in a "superposition" of both 0 and 1 at the same time) have fundamentally unpredictable results.

Again the argument is imprecise:

(1) what does it mean "superior"?
(2) is the answer "because measurements of some quantum particles have fundamentally unpredictable results" good enough?

The first photonic QRNG called *Quantis* was produced by ID Quantique in 2001, and it is based on the standard beamsplitter experiment; see Figures 1 and 2 in [234]. For an experimental analysis of the quality of *Quantis*; see [88, 1, 232, 318].

There are many newcomers, like EvolitionQ[63] and QUANTUM RNG.[64] Most companies justify the quality of their quantum random bits with arguments similar to those of ID Quantique and NIST, discussed above. Cambridge Quantum Computing was a UK-based quantum software startup[65]

[62] `phys.org/news/2018-04-quantum-method-random.html`.

[63] `evolutionq.com/quantum-safe-publications/qrng-report-2021-evolutionQ.pdf`.

[64] `bilgem.tubitak.gov.tr/en/urunler/quantum-rng-quantum-based-random-numbe r-generator`.

[65] CQC became part of the company Quantinuum `www.quantinuum.com/about`; it seems the quantum randomness was dropped as of 13 March 2023.

that sold "certifiably quantum, cloud-based random number generation service":[66]

> "Randomness is a global commodity – people pay for it by the bit," says Ilyas Khan, chief executive and founder of CQC, in an interview with Sifted.
>
> The problem is that genuine, verifiable randomness – think of it as the property possessed by a sequence of numbers that makes it impossible to predict the following number in the sequence – is extremely hard to come by. You can simulate randomness in various ways – by sampling atmospheric noise, for example – but generally speaking, classical computers are not good at doing this.
>
> Quantum computers, on the other hand, can generate genuinely random numbers.

CQC went one step further than the other companies: they used Bell's test[67] to certify the quantum random bits produced by a quantum computer, used randomness extractors to "obtain perfect randomness" and sell these quantum random bits over the Internet. "A second line of business will sell cryptography keys that are guaranteed to be quantum-proof." Incredible, isn't it? It looks too good to be true. First, the reader remembers that "perfect randomness" does not exist; second, there is no proof that the quality of randomness is better than that of those produced by PRNGs; third, there is no explicit protocol to guarantee that the quantum random bits sold are "genuinely" created with the quantum computer and not with a very good PRNG!

Can we do it better? One way to do this is to use a "located" form of the Kochen-Specker Theorem, a deterministic no-go theorem in quantum mechanics. The Kochen-Specker Theorem [249] is valid only in Hilbert spaces of dimension of at least 3, hence the certification given in what follows does not work for the 2-dimensional QRNGs [219], in particular for the beamsplitter used by Quantis [233].

Next, we present the necessary notions and assumptions of quantum theory, followed by the main result, which allows the "algorithmic location" of value indefinite observables. A *definite value* is precisely a (deterministic) hidden variable specifying, in advance, the result of the measurement of an observable.

[66]`sifted.eu/articles/finally-a-way-to-make-money-out-of-quantum-selling-randomness/`.

[67]Bell Theorem is a probabilistic no-go theorem [38, 298].

We use the standard notation [283] in quantum physics. The complex n-dimensional space is denoted by $\mathbf{C}^n$ ($\mathbf{C}$ is the set of complex numbers). A ket $|v\rangle$ denotes a vector (state) in $\mathbf{C}^n$. A bra $\langle f|$ denotes a linear function $f\colon V \to \mathbf{C}^n$ that maps each vector in space V to an element in $\mathbf{C}^n$. The linear function $\langle f|$ on the vector $v\rangle|$ is written $\langle f|v\rangle \in \mathbf{C}^n$. We denote the observable projecting onto the linear subspace spanned by a vector $|\psi\rangle \in \mathbf{C}^n$ as $P_\psi = \frac{|\psi\rangle\langle\psi|}{|\langle\psi|\psi\rangle|}$.

We then fix a positive integer $n > 2$ and let $O \subseteq \{P_\psi \mid |\psi\rangle \in \mathbf{C}^n\}$ be a nonempty set of one-dimensional projection observables on the Hilbert space $\mathbf{C}^n$. See more in [300].

The quantum measurement of an observable is non-contextual if its outcome is independent of the "context", i.e. is independent of how the observable is measured. The formal definitions are as follows. A set $C \subset O$ is a *context* of O if C has n elements and for all $P_\psi, P_\phi \in C$ with $P_\psi \neq P_\phi$, $\langle\psi|\phi\rangle = O$. A *value assignment function* (on O) is a partial function $v : O \to \{0, 1\}$ assigning values to some (possibly all) observables in O.[68] An observable $P \in O$ is *value definite* (under the assignment function v) if $v(P)$ is defined; otherwise, it is *value indefinite* (under v). Similarly, we call O *value definite* (under v) if every observable $P \in O$ is value definite.

We assume the following hypotheses:

- **Admissibility:** Let O be a set of one-dimensional projection observables on $\mathbf{C}^n$ and let $v : O \to \{0, 1\}$ be a value assignment function. Then v is *admissible*[69] for O if for every context C of O, we have $\sum_{P\in C} v(P) = 1$, i.e. only one projection observable in a context can be assigned the value 1.
- **Non-contextuality of definite values**: The outcome obtained by measuring a value definite observable (a pre-existing physical property) is *non-contextual*, i.e. it does not depend on other compatible observables which may be measured alongside it.

Value indefinite observables are essential because, as we will show, *measuring one such observable produces a "random" outcome.* To measure a value indefinite observable, we have to "effectively find" one, so the existential Kochen-Specker Theorem is not enough.[70] Motivated by Einstein, Podolsky and Rosen's definition of *physical reality* [171, p. 777]:

[68]The partiality of the function v means that $v(P)$ can be $0, 1$ or indefinite.
[69]That is, in agreement with quantum mechanics predictions.
[70]Even in case the finite set has two elements.

> If in any way, without disturbing a system, we can predict with certainty the value of a physical quantity, then there exists a *definite value* before observation corresponding to this physical quantity.

we adopt the following [8]:

- **Eigenstate principle**: If a quantum system is prepared in the state $|\psi\rangle$, then the projection observable P_ψ is value definite.

In detail, if a quantum system is prepared in an arbitrary state $|\psi\rangle \in \mathbf{C}^n$, then the measurement of the observable P_ψ should yield the outcome 1, hence, if $P_\psi \in O$, then $v(P_\psi) = 1$.

Located Kochen-Specker Theorem. [2, 5, 8] *Consider a quantum system described by the state $|\psi\rangle$ in a Hilbert space $\mathbf{C}^n$, $n \geq 3$. Choose a state $|\phi\rangle$ that is neither orthogonal nor parallel to $|\psi\rangle$ $(0 < |\langle\psi|\phi\rangle| < 1)$. If the following three conditions are satisfied: i) admissibility, ii) non-contextuality and iii) eigenstate principle, then the projection observable P_ϕ is value indefinite.*

Next, we present a setup satisfying the conditions of Located Kochen-Specker Theorem that guarantees the value indefiniteness of the observables, does not rely on probabilistic results, and *ensures maximal unpredictability and robustness to errors* (as in the case of multiple photon emissions).

To fulfil the Hilbert space dimensional requirement, we can use a collection of theoretical beamsplitters representing the state of a spin-1[71] particle [2] as described by its corresponding unitary decomposition, where the desired probability distribution can be achieved with careful state preparation.

It is mathematically useful to work with the following variant of the Located Kochen-Specker Theorem:

Spectral Located Kochen-Specker Theorem. [15] *Let O be an observable with spectral decomposition $O = \sum_{i=1}^{n} \lambda_i P_{\lambda_i}$, where λ_i denotes each distinct eigenvalue with corresponding eigenstate $|\lambda_i\rangle$. Then, O has a predetermined measurement outcome if and only if each projector in its spectral decomposition has a predetermined measurement outcome.*

[71] Many results in this section hold for an arbitrary 3-dimensional particle.

The main effort in what follows is to represent the 3-dimensional unitary operator required by the Kochen-Specker Theorem as a composition of 2-dimensional operators preserving the value indefiniteness.

The Spectral Located Kochen-Specker Theorem works for the outcome measurements of observables with non-degenerate spectra. Furthermore, let $C = \{P_1, \ldots, P_n\}$ be a context, i.e. a maximal set of compatible projection observables, and let v be a value assignment function such that $v(P_1) = 1$ under C. If any pair (P_1, P_i) is measured, then the system will collapse into the eigenstate $|\phi\rangle$ of the projection observable P_1 with eigenvalue 1. As all observables in C are physically co-measurable and $\sum_{j=1}^{n} P_j = 1$, we deduce that $|\phi\rangle$ is an eigenstate of P_i with corresponding eigenvalue 0, hence $v(P_i) = 0$. Similarly, if $v(P_i) = 0$ for all $i \neq 1$, then $v(P_1) = 1$. Hence, the admissibility property of v serves as a generalisation of the sum rule that corresponds to the physical interpretation of the measurement process.

The property denoted by *spin* ($\mathbf{S}$) is the intrinsic form of angular momentum characteristic of elementary particles. By deriving the spin state operator S_x, we can analyse the preparation state $|S_z\rangle$ effect on the outcome probabilities. We consider the description of states that point in arbitrary directions specified by the unit vector $\mathbf{u} = (u_x, u_y, u_z) = (\sin\theta\cos\vartheta, \sin\theta\sin\vartheta, \cos\theta)$, where θ, ϑ are the polar and azimuthal angles; we then define the spin observable operator $\mathbf{S}$ as a triplet of operators $\mathbf{S} = (S_x, S_y, S_z) = \hbar\boldsymbol{\sigma}$, where $\boldsymbol{\sigma}$ corresponds to the generalised Pauli matrices for a spin-1 particle. Then, by adopting units in which $\hbar$ is numerically equal to unity, we obtain the generalised observable that describes the measurement context:

$$S(\theta, \vartheta) = \mathbf{u} \cdot \mathbf{S} = \begin{pmatrix} \cos(\theta) & \frac{e^{-i\vartheta}\sin(\theta)}{\sqrt{2}} & 0 \\ \frac{e^{i\vartheta}\sin(\theta)}{\sqrt{2}} & 0 & \frac{e^{-i\vartheta}\sin(\theta)}{\sqrt{2}} \\ 0 & \frac{e^{i\vartheta}\sin(\theta)}{\sqrt{2}} & -\cos(\theta) \end{pmatrix}. \tag{4.2}$$

Using the orthonormal Cartesian standard basis $|1\rangle = (1, 0, 0), |0\rangle = (0, 1, 0)$ and $|-1\rangle = (0, 0, 1)$, and the eigenvalues $\{-1, 0, 1\}$ of S_x we obtain the unitary matrix U_x corresponding to the spin state operator S_x:

$$U_x = \frac{1}{2}\begin{pmatrix} 1 & \sqrt{2} & 1 \\ \sqrt{2} & 0 & -\sqrt{2} \\ 1 & -\sqrt{2} & 1 \end{pmatrix}. \tag{4.3}$$

In particular, S_z is given by $S(0, 0)$ and S_x by $S(\frac{\pi}{2}, 0)$.

As demonstrated in [335], given an arbitrary unitary operator, we can represent a generalised rotation through the decomposition of the unitary matrix U_x using a series of phase shifters and beamsplitters implemented in an optical experiment.

As unitary decompositions are not unique, the unavoidable imperfections in every experimental setup mean that only some choices are suitable for physical implementation. Consequently, a unitary decomposition must be carefully constructed to reduce internal loss, minimise the physical footprint, and make the implemented transformation as close as possible to the ideal one.

The calculation in [15] proposed the decomposition:

$$U_x = B_{1,2}^{-1} \cdot B_{2,3}^{-1} \cdot D \cdot B_{1,2} = B_{1,2}^{-1} \cdot B_{2,3} \cdot D \cdot B_{1,2}, \tag{4.4}$$

where

$$B_{1,2} = \begin{pmatrix} \sqrt{\frac{2}{3}} & \frac{1}{\sqrt{3}} & 0 \\ \frac{i}{\sqrt{3}} & -i\sqrt{\frac{2}{3}} & 0 \\ 0 & 0 & 1 \end{pmatrix}, \; B_{2,3} = \begin{pmatrix} 1 & 0 & 0 \\ 0 & \frac{1}{2} & -\frac{i\sqrt{3}}{2} \\ 0 & \frac{i\sqrt{3}}{2} & -\frac{1}{2} \end{pmatrix} = B_{2,3}^{-1},$$

$$B_{1,2}^{-1} = \begin{pmatrix} \sqrt{\frac{2}{3}} & -\frac{i}{\sqrt{3}} & 0 \\ \frac{1}{\sqrt{3}} & i\sqrt{\frac{2}{3}} & 0 \\ 0 & 0 & 1 \end{pmatrix}, \; D = \begin{pmatrix} 1 & 0 & 0 \\ 0 & -1 & 0 \\ 0 & 0 & -1 \end{pmatrix}.$$

To justify the use of the *2-dimensional* matrices representing beamsplitters to construct the *3-dimensional* unitary operator, we have to prove that the 2-dimensional decomposition induces a mapping that preserves the *3-dimensionality*, hence value indefiniteness, so the Located Kochen-Specker Theorem applies; this is essential as this theorem is false in dimension two.

Invariance Theorem. **[15]** *The operator U_x defined by (4.4) preserves 3-dimensionality, hence value indefiniteness, i.e. if ψ is value indefinite, then $U_x(\psi)$ is also value indefinite.*

Next, we construct a value indefinite quantum state, which by measurement, produces outcomes with a given probability distribution (p_1, p_2, p_3) where $p_1 + p_2 + p_3 = 1$ and $0 < p_i < 1$.

Value Indefinite Quantum States Theorem. [15] *The following quantum states are value indefinite with respect to the standard Cartesian basis:*

$$\begin{aligned}|\phi\rangle^* &= \left[\pm\sqrt{2}\sqrt{p_2} + z\right]|1\rangle \\ &\quad + \left[\pm\sqrt{p_2} \mp \sqrt{2}\sqrt{p_3} + z\sqrt{2}\right]|0\rangle \\ &\quad + \left[\pm\frac{\sqrt{p_1}}{2} \mp \frac{\sqrt{p_2}}{\sqrt{2}} \pm \frac{\sqrt{p_3}}{2}\right]|-1\rangle .\end{aligned} \tag{4.5}$$

for every combination of the signs + and −.

Accordingly, given a probability distribution (p_1, p_2, p_3), every quantum state in (4.5) is a valid preparation state for the 3D-QRNG, and this is obtained by choosing a combination of signs for $|\phi^*\rangle$.

For the probability distribution $(\frac{1}{4}, \frac{1}{2}, \frac{1}{4})$, by setting

$$(+\sqrt{p1}, +\sqrt{p2}, +\sqrt{p3}) = \left(\frac{1}{2}, \frac{1}{\sqrt{2}}, \frac{1}{2}\right),$$

we can obtain the valid preparation state

$$|\phi\rangle = [1+z]\,|1\rangle + \left[\frac{1}{\sqrt{2}} - \frac{\sqrt{2}}{2} + z\sqrt{2}\right]|0\rangle + \left[\frac{1}{4} - \frac{1}{2} + \frac{1}{4}\right]|-1\rangle = |1\rangle .$$

Similarly, for the probability distribution $(\frac{1}{3}, \frac{1}{3}, \frac{1}{3})$, we get the following valid preparation states

$$\pm\frac{1}{\sqrt{3}}\left(|1\rangle + |-1\rangle\right) \pm \frac{1}{\sqrt{6}}\left(|1\rangle - |-1\rangle\right), \frac{1}{\sqrt{6}}|1\rangle \pm \sqrt{\frac{2}{3}}|0\rangle - \frac{1}{\sqrt{6}}|-1\rangle ,$$

$$-\frac{1}{\sqrt{6}}|1\rangle \pm \sqrt{\frac{2}{3}}|0\rangle + \frac{1}{\sqrt{6}}|-1\rangle .$$

Next, we prove that *every sequence produced by the proposed* 3D-QRNGs *is incomputable*; that is, *no sequence produced by such a* 3D-QRNG *can be reproduced exactly by any algorithm, in particular, by any pseudo-random generator. This shows that the quality of the quantum random digits produced by every* 3D-QRNG *described in this paper is provably better than the one made by* any *pseudo-random number generator.*

In detail, consider a process that algorithmically repeats the process of state preparation and measurement, as described above, and let $\mathbf{x} =$

$x_1x_2\ldots$ be the infinite sequence produced by the measurement outputs; here each x_i is 0 or 1 or 2. Let $\mathcal{O}, \mathcal{C}$ be two fixed sets of observables and contexts, whose respective components O_i, C_i denote the observable and the corresponding context of the i-th measurement. Let $f : \mathbf{N} \times \mathcal{O} \times \mathcal{C} \to A_3$ be the function defined by $f(i, O_i, C_i) = x_i$ for every i. The incomputability of $\mathbf{x}$, which is equivalent to the incomputability of f, follows from the non-contextuality of definite values; see [2, Section B] for details. We say that a measurement outcome is predictable if f is *computable*; otherwise, f is incomputable, so it offers no method of prediction [362].

A stronger result can be obtained by using the non-probabilistic model for unpredictability [6, 7]. To this end, we consider an *experiment* E producing a single-digit $x \in A_3$. With a particular trial of E, we associate the parameter λ (the state of the universe), which fully describes the trial; λ is a resource from which one can extract finite information to predict the outcome of the experiment E. The trials of E generate a succession of events of the form "E is prepared, performed, the result is recorded, E is reset", algorithmically iterated finitely many times.

An *extractor* is a physical device selecting a finite amount of information from λ without altering the experiment E; the outcome is a string of digits $\langle\lambda\rangle$ over A_3. A *predictor* for E is an algorithm P_E which *halts* on every input and *produces* an element of A_3 or *prediction withheld.*

The predictor, P_E, can use the information $\langle\lambda\rangle$ as input but must be *passive,* that is, it must not disturb or interact with E in any way. A predictor P_E provides a *correct prediction* using the extractor $\langle\,\rangle$ for an instantiation of E with parameter λ on the input $\langle\lambda\rangle$, in case it outputs an element of A_3 (that is, it does not refrain from making a prediction) that is equal to x, the result of the experiment.

Fix an extractor $\langle\,\rangle$ and a positive integer k. The predictor P_E is $k, \langle\,\rangle$-*correct* if there exists an $n \geq k$ such that when E is repeated n times with associated parameters $\lambda_1, \ldots, \lambda_n$ and produces the outputs $x_1, x_2, \ldots, x_n$, then P_E outputs the sequence $P_E(\langle\lambda_1\rangle), P_E(\langle\lambda_2\rangle), \ldots, P_E(\langle\lambda_n\rangle)$ with the following two properties: (i) no prediction in the sequence is incorrect, and (ii) in the sequence there are k correct predictions.

If P_E is $k, \langle\,\rangle$-correct, the probability that P_E is operating by chance and may not continue to give correct prediction is bounded by $3^{-n}\binom{n}{k} < \frac{2^n}{3^n} \leq \left(\frac{2}{3}\right)^k$. This probability tends exponentially to 0 when $k \to \infty$, so the confidence we have in a $k, \langle\,\rangle$-correct predictor increases exponentially with k.

If P_E is $k, \langle\,\rangle$-correct for all k, then P_E never makes an incorrect prediction, and the number of correct predictions can be made arbitrarily large by repeating E enough times. If P_E is not $k, \langle\,\rangle$-correct for all k, then we cannot exclude the possibility that every correct prediction P_E makes is simply due to chance. Consequently, we can define the predictability of a single trial:

The outcome x of a single trial of the experiment E performed with parameter λ is *predictable* (with certainty) if there exist an extractor $\langle\,\rangle$ and a predictor P_E which is $k, \langle\,\rangle$-correct for all k, and $P_E(\langle\lambda\rangle) = x$.

In this case, if the predictor P_E outputs x, then P_E never makes an incorrect prediction no matter how many times it is used, practically finitely many, theoretically infinitely many.

Recall that the 3D-QRNG works in a 3-dimensional Hilbert space, so the outcome of each measurement is 0 or 1 or 2, an element of the $A_3 = \{0, 1, 2\}$.

A sequence $\mathbf{x}$ over the alphabet A_3 is called *3-bi-immune* if there is no partial computable function $\varphi : \mathbf{N} \to A_3$ such that its domain is infinite and $\varphi(i) = x_i$ for every i in the domain of φ; see [84]. Every 3-bi-immune sequence is incomputable, but the converse is not true.[72]

Theorem. [15] *A sequence* $\mathbf{x} \in A_3^\omega$ *is 3-bi-immune if and only if no single digit of* $\mathbf{x}$ *can be predicted by any predictor.*

Assuming the **Eigenstate principle**, and the

epr principle. *If a repetition of measurements of an observable generates a computable sequence, then this implies these observables were value definite.*

the following results follow from Theorem 3 in [7]:

Theorem. [15] *a) Let* $\mathbf{x}$ *be an infinite sequence obtained by measuring a quantum value indefinite observable in* $\mathbf{C}^3$ *in an algorithmic infinite repetition of the experiment* E*. Then no single-digit* x_i *can be predicted.*

b) Let $\mathbf{x}$ *be an infinite sequence obtained by measuring a quantum value indefinite observable in* $\mathbf{C}^3$ *in an algorithmic infinite repetition of the experiment* E*, then* $\mathbf{x}$ *is 3-bi-immune.*

[72]Challenge for the reader: Prove this claim.

Using Value Indefinite Quantum States Theorem, every quantum state in (4.5) is value indefinite and measuring it with the universal unitary operator U_x produces a quantum random ternary digit.

Theorem. [15] *Every* 3D-QRNG *that uses a value indefinite observable (4.5) and the universal unitary operator (4.4) always generates sequences for which no single digit can be predicted. In particular, every such sequence is* 3*-bi-immune.*

The following simple morphism $\varphi\colon A_3 \to A_2$ transforms a ternary sequence into a binary sequence:

$$\varphi(a) = \begin{cases} 0, & \text{if } a = 0, \\ 1, & \text{if } a = 1, \\ 0 & \text{if } a = 2, \end{cases} \tag{4.6}$$

which can be extended sequentially for strings, $\mathbf{y}(n) = \varphi(\mathbf{x}(n)) = \varphi(x_1)\varphi(x_2)\ldots\varphi(x_n)$ and sequences $\mathbf{y} = \varphi(\mathbf{x}) = \varphi(x_1)\varphi(x_2)\ldots\varphi(x_n)\ldots$. This transformation preserves 2-bi-immunity:[73]

Theorem. [15] *Assume the Eigenstate and epr principles. Let* $\mathbf{y} = \varphi(\mathbf{x})$*, where* $\mathbf{x} \in A_3^\omega$ *is a ternary sequence generated by the* 3D-QRNG *and* φ *is the alphabetic morphism defined in (4.6). Therefore, no single bit of* $\mathbf{y} \in A_2^\omega$ *can be predicted.*

To recapitulate, every PRNG generates a computable sequence of random digits, hence all digits are algorithmically predictable. In contrast, no single digit is predictable, so the *3D-QRNG is provably better than any PRNG.*

Finally, can the above theory be used to build 3D-QRNGs? The answer is affirmative, and here is a solution. Using (4.4), we observe that $B_{2,3} \cdot D = D \cdot B_{2,3}$, hence we get:

$$U_x = B_{1,2}^{-1} \cdot B_{2,3} \cdot D \cdot B_{1,2} = D' \cdot B'_{1,2} \cdot B_{2,3} \cdot B_{1,2}.$$

These relations allow us to set the reflectivity, transmittance and phase shift values for a physical realisation via Mach-Zehnder interferometers:

Table 4.1, with $\eta = 2\arccos\left(\sqrt{\frac{2}{3}}\right)$, provides the correspondence between equation (4.4) and its physically realisable optical implementation in Figure 4.10.

[73]Challenge for the reader: Define this notion.

$B_{m,n}$	θ	ϑ
$B'_{1,2}$	$-\frac{\eta}{2}$	π
$B_{2,3}$	$\frac{2\pi}{3}$	π
$B_{1,2}$	$\frac{\eta}{2}$	$\frac{3\pi}{2}$

Table 4.1: From equation (4.4) to implementation

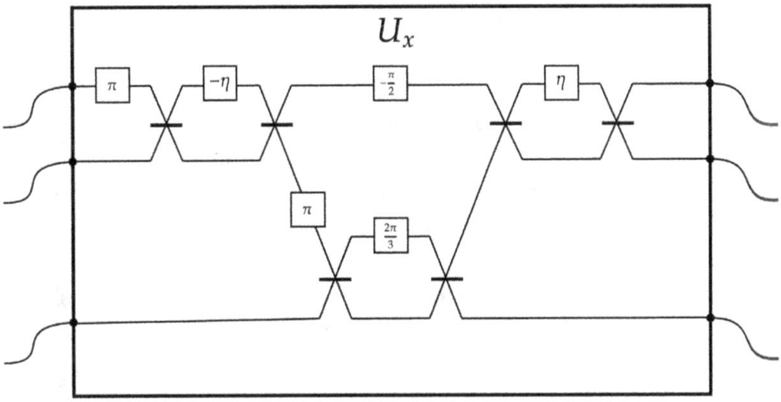

Fig. 4.10: Implementation of the universal unitary decomposition U_x by means of 3-mode multiport interferometer.

The generic arrangement in Figure 4.10 can be realised with multimode interferometers (MMIs). Single-mode waveguides are coupled into a multimode fibre with a certain number of allowed modes. Phase modulation can be achieved by using a thermal phase shifter, which, in silicon devices, can be implemented by connecting a resistive strip of silicon beside the waveguide to a metal pad to which voltage is applied to control the temperature. By integrating two MMIs (acting as balanced beamsplitters) with a thermal phase shifter, a Mach-Zehnder Interferometer with tunable reflectivity implements the arrangement in Figure 4.10.

An arrangement of Mach-Zehnder interferometers consisting of phase shifters and balanced directional couplers illustrates its construction; $\eta = \arccos\left(\sqrt{\frac{2}{3}}\right)$.

4.13 The Halting Problem in Chess and parity game

Halting has practical importance in games in which every match should terminate. One such example is the game of chess. From the very beginning, chess was related to mathematics. The inventor of chess was asked by his King to choose a gift as a reward for his invention. The old man modestly asked the King to give him one grain on the first square; twice as many on the second square, i.e. 2; twice as many on the third square, i.e. $2 \times 2 = 4$; and so on. Indeed, just a few grains of wheat: $1 + 2 + 4 + 8 + \cdots + 2^{64} = 18,446,744,073,709,551,616$.

It had always been the intention that every match should end after finite moves, and the rule that a game is a draw if the same sequence of moves occurs three times in succession was meant to enforce it. Until the third decade of the last century, chess players were not aware of the possibility that a never-ending game of chess could be played according to the rules that were then in force (the draw rule included). In 1929, the Dutch mathematician and chess grandmaster M. Euwe[74] published the paper "Set theoretical considerations about the game of chess" [172], in which he proved that infinite chess matches are indeed legally possible. To this aim, he constructed an infinite sequence of 0s and 1s with no three identical consecutive subsequences of any length and then used it to show that in chess, the draw rule is not sufficient to prohibit infinite games.

In his paper, Euwe independently discovered the sequence now called the Prouhet-Thue-Morse [19, p. 13]

$$0110100110010110100101100110100\,1$$
$$1001011001101001011010011001011\,0\ldots$$

which is obtained by starting with 0 and successively appending the Boolean complement of the sequence obtained thus far:

[74]He was the fifth world chess champion in modern history 1935–37. In 1950, Euwe became interested in computing science and was appointed a Professor of Cybernetics. While visiting the United States to study computing science in 1957, he played two unofficial chess games in New York against the chess prodigy Bobby Fischer, winning one and drawing the second.

$$0, 01, 0110, 01101001, 0110100110010110, \ldots$$

Due to Euwe's discovery, new rules had to be added to the game of chess. B. Fisher, who made a lasting contribution to chess, see, for example, his book [181], was actively involved in searching for new rules to avoid infinite games and stimulate creativity in the game. In 1996 he invented a variation of the game of chess called *Fischer random chess* or *Chess960*, in which the initial position of the pieces is randomised following specific rules to one of 960 possible positions. Randomisation nullifies the advantage given by memorising openings, so players must rely more on their talent and creativity.

In 1948 Turing and his long-time friend D. Champernowne[75] devised a chess-playing program called Turochamp, implemented as a "paper machine" since there was no machine to run it. Turing acted as a "human CPU" requiring more than half an hour per move. In 1952, Turochamp lost to A. Glennie,[76] one of Turing's colleagues.

Parity games form a natural class of games that are not only interesting in their own right, but also because they are connected to fundamental notions like μ-calculus, modal logics, tree automata and Muller games [95].

A parity game is given by a directed graph (V, E), a starting node $s \in V$, and a function *val* which attaches to each $v \in V$ an integer value (also called colour) from a set $\{1, 2, 3, \ldots, m\}$. The main parameter of the game is n, the number of nodes, and the second parameter is m. Two players, Anke and Boris, move alternately in the graph, with Anke moving first. A move from a node v to another node w is valid if (v, w) is an edge in the graph; furthermore, it is required that from every node one can make at least one valid move. Anke starts from the node s.

The alternate moves by Anke and Boris define a finite or an infinite sequence of nodes which is called a *play*. Each value is owned by one player; without loss of generality one player owns the odd numbers and the other player owns the even numbers.

The winner of a finite play is the player whose opponent is unable to move. In an infinite play, Anke wins a play through nodes $a_0, a_1, a_2, \ldots$ if the largest value appearing infinitely often of the sequence $val(a_0), val(a_1), val(a_2), \ldots$ is a number she owns, that is, a number of her

[75]See Section 5.9.

[76]The British computer scientist famous for developing the Autocode, arguably the first computer compiler.

parity. Similarly, Boris wins a play if the largest infinitely often visited node value is owned by him.

An example of a parity game is given in Fig. 4.11.

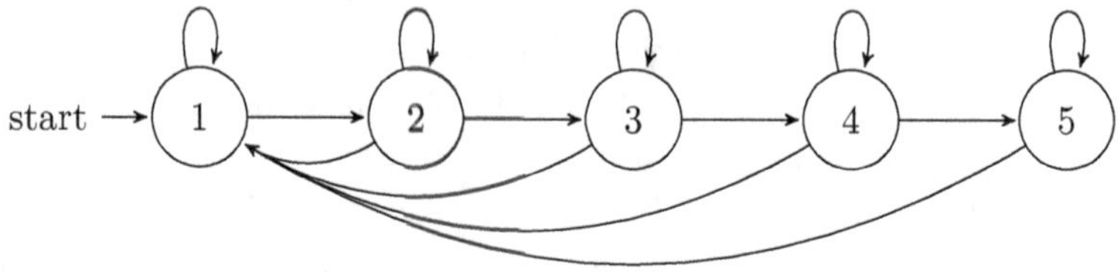

Fig. 4.11: An example of parity game

The nodes are labelled with their values, which are unique (not obligatory); furthermore, Anke has even, and Boris has odd parity. Boris has now the following memoryless (that is, moves are independent of the history) winning strategy for this game: $1 \to 1$, $2 \to 3$, $3 \to 3$, $4 \to 5$, $5 \to 5$. Whenever the play leaves node 1 and Anke moves to node 2, then Boris will move to node 3. If Anke moves to node 4, Boris will move to node 5. Hence, whenever the play is in a node with an even value (this only happens after Anke moved it there), the play will go into a node with a higher odd value in the next step. So the largest infinitely often visited node value is odd, which justifies Boris' win. Hence Boris has a winning strategy for the parity game given above.

Solving a parity game means deciding, for a given starting position, which of the two players has a winning strategy.

The difficulty of solving the parity game comes from the infinity of some plays. Is solving the parity game undecidable? The satisfiability problem for monadic second-order logic – a fragment of the second-order logic where the second-order quantification is limited to monadic predicates, i.e. predicates having a single argument – is undecidable because this logic subsumes first-order logic (or predicate logic), a result intimately related to the Halting problem and established independently by Church [125] and Turing [397]. Contrary to intuition suggesting an undecidability result, Rabin [331] showed that there is an algorithm solving an arbitrary parity game.

Due to the extensive importance of parity games in other areas of mathematics and computing science, investigating the game's complexity was, and continues to be, a challenging problem. The framework for such a study assumes that the game is given by a description in polynomial size in the number n of nodes and that one can evaluate all relevant parts of the description in logarithmic space. A natural possibility is to store the following three items for each game (where Anke moves first and starts from node 1):

(1) two numbers m, n with $1 \leq m \leq n$ and one bit which says whether the values owned by player Anke are the even or the odd numbers;
(2) the game graph given by a table, that is, for each pair of nodes, a bit which says whether there is a directed edge between the two nodes (which can be same) or not;
(3) the values of the nodes given by another table which holds, for each node, a binary number from $\{1, 2, 3, \ldots, m\}$.

An essential question for parity games is the evaluation of the *time complexity* for finding the winner of a parity game when both players play optimally. The first algorithms took exponential time [294, 426], and subsequent studies searched for better algorithms; see the extensive bibliography in [95]. A quasipolynomial time algorithm was reported in [95]; it is an *open problem* whether a polynomial-time algorithm exists for parity games.

4.14 Artificial intelligence: power and limits

Machine learning, deep learning, and neural networks are artificial intelligence (AI) sub-domains. They are not independent fields: deep learning is a sub-field of neural networks, and neural networks are a sub-domain of machine learning. They differ in the way their algorithms "learn". For example, deep learning works with unstructured raw data, like text or images, and it can automatically determine the set of features that distinguish different data categories from one another with no human intervention. Non-deep learning requires more structured data; humans or other algorithms usually produce them. Neural networks are made of node layers – an input layer, one or more hidden layers, and an output layer. Each node connects to another and has an associated weight and threshold. If the output of any individual node is above the specified threshold value, that node is activated, sending data to the next layer of the network; otherwise, no data is passed along to the next layer of the network. A neural network with more than three layers is considered a deep learning algorithm. Deep learning accelerates progress in many areas, such as computer vision, natural language processing, recommendation, stock trading, fraud detection, and speech recognition. A survey by McKinsey Global Institute has estimated that AI analytics could add around $13trn, or 16%, to annual global GDP by 2030. Adding machine learning to existing business processes is a worldwide trend.

Here is an example of a big success. In November 2022, the start-up company Abacus.AI[77] has released the first version of ChatGPT, a free chatbot that can answer all sorts of questions with unprecedented "eloquence".

ChatGPT, which has become very popular instantaneously[78], is an improvement of GPT-3, an AI model that generates text based on patterns fed from massive Internet databases. ChatGPT stands out because it can answer questions in natural languages, a new interface everyone can use. Asking questions in natural language is not new: Siri, Apple's virtual assistant, uses voice queries, gesture-based control, focus-tracking and a natural language to answer questions, make recommendations, and perform actions.

In early February 2023, Google and Baidu unveiled their chatbots called Bard and Ernie, which will compete with ChatGPT.

ChatGPT has been initially used as an alternative to Google because it can provide descriptions, answers and even solutions to complex questions. What does the program do? According to ATRIA Innovation[79]:

(1) With GPT, you can generate coherent and well-written texts in various styles, topics and languages. In addition, news summaries, product descriptions or stories can be generated.
(2) Thanks to this chat, problems can be analysed, and solutions or answers to questions can be generated.
(3) GPT can generate appropriate and consistent responses for a chatbot in various contexts.
(4) It can be used to generate attractive posts and messages for social networks.
(5) With GPT, you can generate reports, e-mails and other content for productivity applications.
(6) Thanks to chat GPT, large data sets can be analysed, and valuable information can be extracted.

---

[77] "Abacus.AI is the world's first AI-assisted data science and end-to-end MLOps platform that enables real-time machine and deep learning at scale for common enterprise use cases. You can bring your models or use our neural network and generative AI techniques to create highly-accurate models and operationalise them in production." cf. `abacus.ai`.

[78] Being free and sometimes producing fun errors helped.

[79] `www.atriainnovation.com/en/how-does-chat-gpt-work/`, January 2023.

How good is it? How does it do it? And why does it work?

Let's start with the first question. S. Altman, OpenAI, co-founder and CEO, believes that "Soon you will be able to have helpful assistants that talk to you, answer questions, and give advice. Later you can have something that goes off and does tasks for you. Eventually, you can have something that goes off and discovers new knowledge for you."[80]

For Stanford University Professor Christopher Potts, ChatGPT is "extremely impressive".[81] Many others agree, but not all.

For Mike Pearl, "The ChatGPT chatbot from OpenAI is amazing, creative, and totally wrong. Need ideas? Great! Need facts? Stay away!"[82] For Professor G. N. Smith from Pomona College, "GPT-3's struggles with reality".[83] E. Musk, the startup co-founder, left the board and distanced himself from the company. The company itself acknowledged that at this stage,[84] ChatGPT can give wrong answers and present misinformation as fact, writing "plausible-sounding but incorrect or nonsensical answers".

To give a short answer to the second question, we refer to a documented article by S. Wolfram [413], which "looks inside" ChatGPT:

> My purpose here is to give a rough outline of what's going on inside ChatGPT-and then to explore why it is that it can do so well in producing what we might consider to be meaningful text. I should say at the outset that I'm going to focus on the big picture of what's going on – and while I'll mention some engineering details, I won't get deeply into them. (And the essence of what I'll say applies just as well to other current "large language models [LLMs] as to ChatGPT.)"

ChatGPT's primary objective is to generate a "plausible follow-up" to any text it receives. But what is "plausible"? Synonyms are "acceptable", "adequate", and "passable". In this context, a "plausible follow-up" could be a sequence of words that aligns with what a human familiar with many relevant sources would expect. According to [415], given the input,

> The best thing about AI is its ability to

produce a ranked list of words that might follow, each with some probabil-

[80] `tinyurl.com/28ypucjc`.
[81] `tinyurl.com/8725fe6u`.
[82] `https://mashable.com/article/chatgpt-amazing-wrong`.
[83] See `http://economics-files.pomona.edu/GarySmith/audiofiles/OnPoint.mp3`.
[84] December 2022.

ity: "learn" (4.5%), "predict" (3.5%), "make" (3.2%), "understand" (3.1%), "do" (2.9%).

Simplifying, this is what ChatGPT does all the time: ask over and over again

> given the text so far, what should the next part of a word (or even word) be?

and based on the answers (and probabilities), decide the adding (part of the) word. Quoting again [413]

> One might think it should be the "highest-ranked" word (i.e. the one to which the highest "probability" was assigned). But this is where a bit of voodoo begins to creep in. Because for some reason – that maybe one day we'll have a scientific-style understanding of – if we always pick the highest-ranked word, we typically get a very flat essay, that never seems to "show any creativity" (and even sometimes repeats word for word). But if sometimes (at random) we pick lower-ranked words, we get a "more interesting" essay.

The key word is *randomness*: if we use the exact random prompt multiple times, we will get different essays each time! There is no underlying theory: empirically, this is the way to produce diversity.

If you are surprised, think about "generative music". The American composer and professor L. Hiller Jr. and mathematician L. Isaacson at Illinois at Urbana-Champagne, used the ILLIAC computer to "compose" music in the early 1950s.

First, they used simple rules: a melody with only notes within an octave, harmonies that tended towards the major and the minor with no dissonance, and a few more parameters. The University's student orchestra members performed compositions sounding like Bach and Penderecki. Still, they didn't sound "original". Eventually, the ingredient which helped was the random insertions of dissonances.[85]

In 1954, they collaborated on *String Quartet No. 4, Illiac Suite*, the first significant use of a computer to compose music. In 1959, they wrote the first book on this topic [224].

Generalising the context, AI has undoubtedly made life easier. Implementing AI has also raised several concerns and questions. Here are some of

[85] For details, see `www.musicradar.com/how-to/ordered-chaos-a-guide-to-randomisation-probability-and-generative-music`.

them: Is the power of AI limitless? Is a form of AI surpassing human intelligence shortly?[86] Is AI understandable? Is AI ethical? Is AI accountable? Is AI free of bias and discrimination? Is AI safe and private?

In what follows, we will discuss only the first question: Is the power of AI limitless? If by limitless, we mean "AI can solve any problem", then the reader would suspect the answer is "no". Why? AI ultimately consists of complex algorithms, whose limits have been proven by computability and complexity theories [311, 60, 61, 418]. In a nutshell:

Theorem. *Most problems cannot be solved by algorithms.*

AI relies on algorithms, hence can "solve" only problems which cannot fall into the class covered by the undecidability theorems. The human brain can write and maintain massively complex software but cannot solve undecidable problems.

There are many articles regarding ChatGPT's positive features – which include assisting education by making reference lists, generating "first drafts", debugging, and tutoring – and adverse side-effects, like being used in exam cheating. As G. Smith noted,[87] the undesirable use of ChatGPT in exam cheating has a positive educational consequence:

> the best response by educators is to teach and test critical thinking skills. These are what students need and they can't be reliably faked by large language models (LLMs).

How does ChatGPT perform in mathematics? The article [185] investigates

> mathematical capabilities of ChatGPT by testing it on publicly available datasets, as well as hand-crafted ones, and measuring its performance against other models trained on a mathematical corpus, such as Minerva. We also test whether ChatGPT can be a useful assistant to professional mathematicians by emulating various use cases that come up in the daily professional activities of mathematicians (question answering, theorem searching).

[86] "Any intellect that vastly outperforms the best human brains in practically every field, including scientific creativity, general wisdom, and social skills", is according to philosopher Nick Bostrum, a *superintelligence*. See also [361, 360].

[87] `mindmatters.ai/2023/03/text-generators-education-and-critical-thinking-an-update/`.

The study concludes

> that contrary to many positive reports in the media (a potential case of selection bias), ChatGPT's mathematical abilities are significantly below those of an average mathematics graduate student. Our results show that ChatGPT often understands the question but fails to provide correct solutions. Hence, if your goal is to use it to pass a university exam, you would be better off copying from your average peer!

Next, we present a simple conversation testing ChatGPT's ability to deal with the infinite. The test is inspired by a conversation of S. Marcus[88] with a young Spanish girl:[89]

Fig. 4.12: S. Marcus[90]

> Antonia, a 10-year-old Romanian-Spanish girl from Tarragona, Spain, had a decisive influence in choosing the topic of the volume. Earlier than us, she called Professor Marcus El infinito, under the

[88]The author's professor, supervisor, mentor, collaborator and friend for more than four decades.

[89]See the Preface of the book [102].

[90]Source: In Wikipedia. `en.wikipedia.org/wiki/File:C14_7404-SolomonMarcus.edited.jpg`.

circumstances pointing to his manner of approaching kids. In his opinion, a child should rediscover Zeno's paradoxes via a Socrates-type dialogue of the following kind: S.M.: "Look, Antonia, you have some usual bread, and you eat half of it today. How much does it remain for tomorrow?" A.: "Half a bread, of course." S.M.: "Okay, but tomorrow you eat half of what you have. Does some bread remain for after tomorrow?" A.: "Yes, sir, some bread still remains." S.M.: "Good, but after tomorrow you again eat half of your bread. Is it true that some bread remains?" A. (already doubtful): "Yes, but" ... After about five iterations, the two parts are always separated: A., or any other "victim", knowing for sure that (s)he can eat a normal bread in a day or so, hence any continuation of the dialogue is senseless (because it's breadless), S.M. looking desperately for the bright light of infinity in the others' eyes: eating every day half of the piece of bread you have got means securing bread for infinitely many days. In many cases, both sides are disappointed ... In at least one case, S.M. got a surname: *El Infinito*.

Below is the dialogue with ChatGPT:

Hello ChatGPT!

Hello! How can I help you today?

You have a loaf of bread, and Antonia eats today half of it. How much does it remain for tomorrow?

If Antonia eats half of the loaf of bread today, then half of the loaf remains for tomorrow.

OK, the next day, Antonia eats half of what remains. Does some bread remain for the following day?

If Antonia eats half of what remains from the previous day, then the remaining bread will be halved again. If the original loaf of bread is divided in half, and then half of the remaining bread is eaten, then one quarter of the original loaf will remain.

OK, the next day, Antonia eats half of what remains. Does some bread remain for the following day?

If Antonia eats half of what remains from the previous day, then the remaining bread will be halved again. If one quarter of the original loaf remains and half of that is eaten, then 1/8th of the original loaf will remain.

OK, the next day, Antonia eats half of what remains. Does some bread remain for the following day?

Yes, if Antonia continues to eat half of what remains each day, some bread will remain for the following day. The amount of bread remaining will decrease by half each day, but it will never reach zero as long as Antonia only eats half of what remains each day.

After clicking "regenerate response" we got:

Hello ChatGPT!

Hello! How can I help you today?

You have a loaf of bread, and Antonia eats today half of it. How much does it remain for tomorrow?

If Antonia eats half of what remains from the previous day, then the remaining bread will be halved again. If 1/8th of the original loaf remains and half of that is eaten, then 1/16th of the original loaf will remain.

OK, the next day, Antonia eats half of what remains. Does some bread remain for the following day?

If Antonia eats half of what remains from the previous day, then the remaining bread will be halved again. If 1/16th of the original loaf remains and half of that is eaten, then 1/32th of the original loaf will remain. And so on, following this pattern, the amount of bread will be halved every day.

OK, the next day, Antonia eats half of what remains. Does some bread remain for the following day?

If Antonia eats half of what remains from the previous day, then the remaining bread will be halved again. If 1/32th of the original loaf remains and half of that is eaten, then 1/64th of the original loaf will remain. The pattern will continue as long as the bread is not completely finished.

OK, the next day, Antonia eats half of what remains. Does some bread remain for the following day?

If Antonia continues to eat half of what remains from the previous day, the amount of bread will continue to be halved each day. However, it will eventually run out, and there will be no bread left for future days.

The ChatGPT answers in the initial and regenerated dialogues are different in many respects, but the striking points are the conclusions:

(1) In the first case:
Yes, if Antonia continues to eat half of what remains each day, some bread will remain for the following day. The amount of bread remaining will decrease by half each day, but it will never reach zero as long as Antonia only eats half of what remains each day.
(2) In the second case:
If Antonia continues to eat half of what remains from the previous day, the amount of bread will continue to be halved each day. However, it will eventually run out, and there will be no bread left for future days.

The last answer in the first dialogue

> Yes, if Antonia continues to eat half of what remains each day, some bread will remain for the following day. The amount of bread remaining will decrease by half each day, but it will never reach zero as long as Antonia only eats half of what remains each day.

seems impressive because instead of responding "exactly" to the question

> OK, the next day, Antonia eats half of what remains. Does some bread remain for the following day?

which refers to the *particular case 1/8th*, it offers a *generalisation* and, based on it, the correct answer. Could it be that ChatGPT was showing a form of mathematical "thinking"? As ChatGPT does not rely on logical or mathematical inferences, is it a kind of emergent[91] skill? The first conclusion is not only correct but also includes a justification, if not a proof.[92]

After "regenerate", ChatGPT seems to again make a more general correct inference

> The pattern will continue as long as the bread is not completely finished.

However, unexpectedly, it ends with a wrong answer.

The first conclusion is correct, but the second is wrong. Why? One can speculate that in the first case, ChatGPT was following a "mathematical" approach, while it offered a more "practical" answer in the second case. A more plausible explanation is that ChatGPT dialogues have been processed

[91]See also [109].
[92]Does this answer suggest the possibility of humanity running out of bread?

independently, despite using the command “regenerate”: quite surprisingly, ChatGPT processed the regenerated sequence of questions as a new dialogue[93] and made no attempt to make any connection between the *same questions asked.* Consequently, it didn’t realise the contradictory final answers and didn’t “see” that the first answer was correct while the second was wrong. This “behaviour” seems to be characteristic of (current?) AI: the inability to make logical connections, hence recognising “true” vs. “false” answers and also to “learn from mistakes”.

Finally, I used the cloud-based AI-enabled typing assistant Grammarly[94] to check ChatGPT’s answers. Grammarly found many issues. For example, Grammarly suggested changing “Yes, if Antonia continues to eat half of what remains each day, some bread will remain for the following day. The amount of bread remaining will decrease by half each day, but it will never reach zero as long as Antonia only eats half of what remains each day.” to “Yes, if Antonia continues to eat half of what remains each day, some bread will stay for the following day. The remaining bread will decrease by half daily but never reach zero if Antonia only eats half of what remains daily.”

Under the title *Introducing ChatGPT* the opening of the Open.AI website[95] says:

> We’ve trained a model called ChatGPT which interacts in a conversational way. The dialogue format makes it possible for ChatGPT to answer follow-up questions, admit its mistakes, challenge incorrect premises, and reject inappropriate requests.

The example discussed above only confirms some claims. ChatGPT and similar AI applications capture a “statistical likelihood relationship” in impressive but imperfect ways. How far can they go?

(1) Could they master syntactical questions correctly?
(2) What about semantical ones?
(3) Could they self-correct?

More futuristic and much-discussed questions are: Do AI systems like

[93]The command “regenerate” should signal repetition.

[94]“Grammarly’s team of computational linguists and deep learning engineers designs cutting-edge algorithms that learn the rules and hidden patterns of good writing by analyzing millions of sentences from research corpora.” See `www.grammarly.com/blog/how-does-grammarly-work/`.

[95]`openai.com/blog/chatgpt`, checked on 12 March 2023.

ChatGPT have human-level of intelligence? Will AI-powered systems surpass human intelligence? Is AI going to be a threat to humanity? We are far from having full answers, so below we present a few recent opinions.

For mathematician M. du Sautoy [166]

> The value of AI might come not so much in making machines that act like humans, but stopping humans acting like machines.

For Y. LeCun, Chief AI Scientist at Meta,

> Current artificial intelligence systems like ChatGPT do not have human-level intelligence and are barely smarter than a dog ...

The argument is based on the fact that ChatGPT is an LLM trained on a massive amount of text, but most of the human knowledge does not come from language. LeCun[96] observed that

> AI can't load a dishwasher, which a 10-year-old could "learn in 10 minutes". ... a five-month-old baby would look at an object floating and not think too much of it. However, a nine-month-year-old baby would look at this item and be surprised, as it realizes that an object shouldn't float.

Mathematician T. Tao has recently written a piece [386] on AI from which we cite the following two paragraphs:

> Current large language models (LLM) can often persuasively mimic correct expert response in a given knowledge domain (such as my own, research mathematics). But as is infamously known, the response often consists of nonsense when inspected closely. Both humans and AI need to develop skills to analyze this new type of text. The stylistic signals that I traditionally rely on to "smell out" a hopelessly incorrect math argument are of little use with LLM-generated mathematics. Only line-by-line reading can discern if there is any substance. Strangely, even nonsensical LLM-generated math often references relevant concepts. With effort, human experts can modify ideas that do not work as presented into a correct and original argument. The 2023-level AI can already generate suggestive hints and promising leads to a working mathematician

[96]`tinyurl.com/h7rd35p3`.

> and participate actively in the decision-making process. When integrated with tools such as formal proof verifiers, internet search, and symbolic math packages, I expect, say, 2026-level AI, when used properly, will be a trustworthy co-author in mathematical research, and in but many other fields as well.

and

> Then what? That depends not just on the technology, but on how existing human institutions and practices adapt. How will research journals change their publishing and referencing practices when entry-level math papers for AI-guided graduate students can now be generated in less than a day and with the far better accuracy of future AI tools? How will our approach to graduate education change? Will we actively encourage and train our students to use these tools?
>
> We are largely unprepared to address these questions. There will be shocking demonstrations of AI-assisted achievement and courageous experiments to incorporate them into our professional structures. But there will also be embarrassing mistakes, controversies, painful disruptions, heated debates, and hasty decisions.

The authors of the article "The Curse of Recursion: Training on Generated Data Makes Models Forget" [358] rightly noted that "Stable Diffusion revolutionised image creation from descriptive text. GPT-2, GPT-3(.5) and GPT-4 demonstrated astonishing performance across a variety of language tasks". But

> What will happen to GPT-n once LLMs contribute much of the language found online? We find that use of model-generated content in training causes irreversible defects in the resulting models, where tails of the original content distribution disappear. We refer to this effect as Model Collapse and show that it can occur in Variational Autoencoders, Gaussian Mixture Models and LLMs. ... the value of data collected about genuine human interactions with systems will be increasingly valuable in the presence of content generated by LLMs in data crawled from the Internet.

At the same time, this trend raises hard questions about ethics, privacy, the legal system, education, and much more. It is equally important to remember that there are always challenges and difficulties in any given period and that humanity has always found ways to overcome them and progress. Individually it is essential to stay informed and engaged in these efforts. One such effort was the open letter "Pause Giant A.I. Experiments",[97] calling for a temporary moratorium on training models larger than G.P.T.-4. The letter had 27,572 signatories (as of 2 May 2023), including the author of this book.

We continue with a more technical issue: the computational limits of the simplest class of AI computational tools, a sub-class of Neural Networks (NNs). This computing paradigm radically differs from conventional computing: NNs can learn how to solve problems through training with big data representing an I/O relation. In this way, empirical data-based knowledge, different from model-based knowledge, is generated from the execution of entirely crafted and understood algorithms. Model-based approaches allow explainability and are amenable to rigorous analysis, while data-based techniques are hard to interpret and understand.

NNs demonstrate remarkable performance in various fields, such as automatic speech recognition, image recognition, natural language processing, drug discovery and toxicology, and automatic game playing. Additionally, they are finding increasingly diverse applications in software and systems engineering.

Fig. 4.13: J. Sifakis[98]

However, despite their impressive practical results, there still needs to be a greater understanding of how machine learning (ML) models achieve their

[97]`https://futureoflife.org/open-letter/pause-giant-ai-experiments/`.
[98]Source: In Wikipedia. `en.wikipedia.org/wiki/File:Joseph_Sifakis_2018.jpg`.

tasks. To address this challenge, Explainable-AI has emerged as a research area that aims to generate models capable of explaining the behaviour of AI-enabled systems [344].

The distinction between data-based and model-based knowledge holds significant importance in various domains, especially in systems engineering, where NNs are integrated into critical systems that can have detrimental consequences if they fail. For example, end-to-end machine learning (ML) solutions in autonomous systems like self-driving cars have sparked heated debates due to the difficulty in accurately assessing their reliability [212, 359]. The current trend is shifting towards intelligent systems that effectively combine data-based and model-based components, leveraging the strengths of each approach while carefully considering the trade-offs between performance and trustworthiness. This trend is further motivated by the intriguing resemblance between these computing paradigms and the two modes of human thinking [240]. Swift, unconscious thinking in humans relies on empirical learning from data, while deliberate, conscious thinking stems from explainable procedural reasoning. The human mind combines these two thinking modes to generate knowledge and solve problems.

Consequently, it is natural to strive for an optimal fusion of these two complementary computing paradigms to tackle the challenge of machine intelligence. Although efforts in this direction are still in their early stages, they are poised to gain momentum soon.

A similar trend is visible in quantum computing and was discussed already: quantum computing itself has less capacity for computing than classical computing (e.g. no quantum machine can simulate a universal Turing machine), and very few quantum algorithms can run faster than any classical counterpart, so a hybrid quantum-classical algorithm is preferable in many cases [18].

It is well understood that data-based empirical learning should be robust to data variations, but guaranteeing this property is a non-trivial problem. Intuitively, robustness can be conceived as a metric property of the learned data sets such that the meanings of very "close" representations should not significantly differ. While there is some understanding of robustness, there needs to be more comprehension of its invariances and determinants.

If we recognise that NNs work well in some cases but not all, then natural questions arise: What are NNs suitable for? When should we use them, and when not? How significant are the cases when NNs do not work? Are they practically irrelevant or just esoteric matters interesting only from a theoretical point of view? Is it possible to distinguish between problems where the application of NNs is obvious and problems where model-based solutions are more adequate? How can we combine these two types of solutions?

It is worth noting that NNs are often well-suited for effectively classifying vast amounts of robust information, as minor modifications to the input typically have a minimal impact on the classification outcome. This characteristic is particularly advantageous for NNs dealing with sensory information and performing perception functions, such as medical image analysis or face recognition. However, there are specific applications where utilising NNs may not be appropriate. For instance, training an NN to analyse the source code of programs and checking for particular properties, even syntactic ones, is likely impractical. This is because software correctness is highly sensitive to even small changes in the source code.

Moreover, the relationship between software syntax and its meaning, as defined by the operational semantics of the programming language, can be intricate and profound. Similar challenges arise when utilising neural networks (NNs) as monitors for identifying failures in software systems. How confident can we be in the decisions made by these NN-based oracles, especially after extensive training using datasets that distinguish between accepted and non-accepted test sequences? As the sensitivity of NN-based systems to input changes increases, the confidence level in their judgements would likely decrease. Is it feasible to formalise and rigorously validate this phenomenon?

Furthermore, there is a complementary issue concerning the impact of coding information on the complexity of the learning process. Considering the example above, the same program can have numerous semantically equivalent representations at different abstraction levels, such as source code, object code, or even in the form of a transition system for finite state programs. How does the choice of representation type affect the complexity of the learning process? Conversely, when dealing with datasets that possess easily learnable properties, can transforming their representation using unconventional scrambling functions impact robustness and consequently increase the complexity of learning? For instance, if the convexity of datasets is crucial for learning a specific property, what is the complexity associated with representations obtained through encoding methods that undermine their convexity?

Following [94], we present some computational limits of computing Boolean functions with Perceptron Neural Nets (PNNs). We start with specific notation. The set of binary strings of length n is denoted by $\{0, 1\}^n$.

Bits will be denoted by x, y and bit-strings by $\mathbf{x}, \mathbf{y}$: depending on the context we write $\mathbf{x} = (x_1, x_2, \ldots, x_n)$ or $\mathbf{x} = x_1 x_2 \ldots x_n$. The Boolean operations will be denoted by $^{-}$ (negation), $\vee$ (disjunction) and $\cdot$ (conjunction). The set of reals is denoted by $\mathbf{R}$; a vector of n real-valued components is denoted by $\mathbf{w}$.

Boolean functions of $n > 1$ variables $f : \{0,1\}^n \to \{0,1\}$, will be shortly called, *functions*. The true/false points of a function f are denoted by $T(f) = \{\mathbf{x} \in \{0,1\}^n \mid f(\mathbf{x}) = 1\}$ and $F(f) = \{\mathbf{x} \in \{0,1\}^n \mid f(\mathbf{x}) = 0\}$, respectively. A function f will be represented in *Full Disjunctive Normal Form (DNF)* [144, p. 123]. The number of true points of f will be denoted by $\#T(f)$; $\#F(f) = 2^n - \#T(f)$. By d, we denote the *Hamming distance* between strings of length n: $d(\mathbf{x}, \mathbf{y})$ is the number of positions i such that $x_i \neq y_i$.

Every function $f : \{0,1\}^2 \to \{0,1\}$ can be naturally extended to $n > 2$ variables in the following way: $f_2 = f$ and $f_n : \{0,1\}^n \to \{0,1\}$, $f_n(x_1, x_2, \ldots, x_n) = f_2(f_{n-1}(x_1, \ldots, x_{n-1}), x_n)$. In this way, we get the functions of n variables OR_n, XOR_n but not $XNOR_n$. We denote by R_n^1, $R_n^2 : \{0,1\}^n \to \{0,1\}$ defined by $R_n^1(\mathbf{x}) = OR_n(\mathbf{x})$ and $R_n^2(\mathbf{x}) = x_1$ and by $PARITY_n : \{0,1\}^n \to \{0,1\}$ the function

$$PARITY_n(\mathbf{x}) = \begin{cases} 1, & \textit{if the number of 0s in } \mathbf{x} \textit{ is odd,} \\ 0, & \textit{otherwise.} \end{cases} \tag{4.7}$$

The function $PARITY_n$ is closely related to XOR: for every $n > 1$, $PARITY_n = XOR_n$, for even n and $PARITY_n = \overline{XOR_n}$, for odd n.

If $T(f) = \{111, 100, 001\}$, then $d(\mathbf{x}, \mathbf{y}) = 2$ for all distinct $\mathbf{x}, \mathbf{y} \in T(f)$. The Boolean function $PARITY_3$ has this property, but R_3^1 and R_3^2 do not. For the Boolean functions $PARITY_3$, R_3^1, and R_3^2 we have $T(PARITY_3) = \{000, 011, 101, 110\}$, $T(R_3^1) = \{001, 010, 011, 100, 101, 110, 111\}$, $T(R_3^2) = \{100, 101, 110, 111\}$. A simple computation shows that for $PARITY_3$, $d(\mathbf{x}, \mathbf{y}) = 2$ for all distinct $\mathbf{x}, \mathbf{y} \in T(PARITY_3)$; R_3^1 and R_3^2 do not have this property.

A binary classifier is a function that decides whether an input belongs to a specific set. A *linear threshold computing unit* or *single-layer PNN* [343, p. 7] computes a function $P_{\theta,\mathbf{w}} : \mathbf{R}^n \to \{0,1\}$, depending on two parameters, a threshold $\theta \in \mathbf{R}$ and a vector of n weights $\mathbf{w} = (w_1, w_2, \ldots, w_n) \in \mathbf{R}^n$, defined as follows:

$$P_{\theta,\mathbf{w}}(\mathbf{x}) = \begin{cases} 1, \textit{ if } \sum_{i=1}^{n} w_i x_i \geq \theta, \\ 0, \textit{ otherwise}. \end{cases} \tag{4.8}$$

The functions of the form (4.8) are also called *threshold*.

For every $n > 1$, the functions R_n^1, R_n^2 are threshold functions. First, $R_n^1 = P_{\frac{1}{2},(1,1,\ldots,1)}$ and $R_n^2 = P_{\frac{1}{2},(n,-1,\ldots,-1)}$. From (4.8), for R_n^1 we set $\mathbf{w} = (w_1, w_2, \ldots, w_n) = (1, 1, \ldots, 1)$ and $\theta = \frac{1}{2}$,

$$P_{\theta,\mathbf{w}}(\mathbf{x}) = \begin{cases} 1, \textit{ if } \sum_{i=1}^{n} x_i \geq \frac{1}{2}, \\ 0, \textit{ otherwise}, \end{cases} = R_n^1(\mathbf{x}),$$

and for R_n^2 we set $\mathbf{w} = (w_1, w_2, \ldots, w_n) = (n, -1, -1, \ldots, -1)$ and $\theta = \frac{1}{2}$,

$$P_{\theta,\mathbf{w}}(\mathbf{x}) = \begin{cases} 1, \textit{ if } nx_1 - \sum_{i=2}^{n} x_i \geq \frac{1}{2}, \\ 0, \textit{ otherwise}, \end{cases} = R_n^2(\mathbf{x}).$$

The functions $PARITY_n$, in particular, XOR and $XNOR$, cannot be computed by single-layer PNNs. Indeed, consider the following four vectors in $\{0,1\}^n$: $\mathbf{x}_1 = 0^{n-2}01$, $\mathbf{x}_2 = 0^{n-2}10$, $\mathbf{y}_1 = 0^{n-2}00$, $\mathbf{y}_2 = 0^{n-2}11$. For every n, we have $PARITY_n(\mathbf{x}_1) = PARITY_n(\mathbf{x}_2)$, $PARITY_n(\mathbf{y}_1) = PARITY_n(\mathbf{y}_2)$, $PARITY_n(\mathbf{x}_1) \neq PARITY_n(\mathbf{y}_1)$ and $\mathbf{x}_1 + \mathbf{x}_2 = \mathbf{y}_1 + \mathbf{y}_2$, so the conclusion follows from [23, Theorem 3.7][99].

What makes some functions, but not all, computable by PNNs or even computable by some single-layer PNNs?

This phenomenon is not surprising, but we need to ask the question differently to understand it. First, how can PNNs approximate functions well in practice when the set of possible functions is exponentially larger than the set of practically possible PNNs? Indeed, there are 2^{2^n} different functions of n variables, so a PNN implementing a generic function in this class requires at least 2^n bits to describe, that is, more bits than there are atoms in our universe if $n > 260$ (not a large number of variables for practical applications). A similar analysis points to an exponential difference between the number of different functions of a fixed number of variables (double exponential) and the number of polynomial threshold PNNs.

What is the difference between the functions $PARITY_n$ on one side and the functions R_n^1 and R_n^2 on the other? For $PARITY_n$, a single bit in the input $\mathbf{x}$ can flop the value of $PARITY_n(\mathbf{x})$ from 0 to 1 or

[99] A function f of $n > 1$ variables is a threshold function if for every positive integer k, for every sequence $\mathbf{x}_1, \mathbf{x}_2, \ldots, \mathbf{x}_k \in T(f))$ and every sequence $\mathbf{y}_1, \mathbf{y}_2, \ldots, \mathbf{y}_k \in F(f)$ we have $\sum_{i=1}^{k} \mathbf{x}_i \neq \sum_{i=1}^{k} \mathbf{y}_i$.

vice-versa. In contrast, R^1_n and R^2_n are more robust for variations of their inputs. Indeed, for R^1_n, if two inputs $\mathbf{x}, \mathbf{y}$ contain each an 1, say $x_i = 1$ and $y_j = 1$, then $R^1_n(\mathbf{x}) = R^1_n(\mathbf{y})$, and only $R^1_n(0, \ldots, 0) = 0$; for R^2_n we have $R^2_n(x_1, x_2, \ldots, x_n) = R^2_n(x_1, y_2, \ldots, y_n)$, for all $n > 1, x_1, x_2, \ldots, x_n, y_2, \ldots, y_n \in \{0, 1\}$.

This suggests that the two properties, robustness and sensitivity, could determine PNN's computability or incomputability. To test this hypothesis, we will propose definitions for these properties. Informally, a function is sensitive if a "small variation" in the input will determine a jump in the values of the function from 0 to 1 or vice-versa; a function which is not sensitive is robust. Quantifying the "small variation" will refine the definition.[100]

Let f be a function of $n > 1$ variables. Following [239, p. 57], the sensitivity of f at $\mathbf{x} \in \{0, 1\}^n$ is the number $s(f, \mathbf{x})$ of $\mathbf{y} \in \{0, 1\}^n$ that differ from $\mathbf{x}$ in exactly one bit and satisfy $f(\mathbf{x}) \neq f(\mathbf{y})$. *The sensitivity $s(f)$ of f is the maximum of $s(f, \mathbf{x})$ taken on all $\mathbf{x} \in \{0, 1\}^n$*. We say that f is *fully sensitive* if $s(f) = n$, i.e. $s(f)$ is maximum. Note that fully sensitive Boolean functions f exist that are constant (robust) on as many inputs as we wish up to $2^n - n - 1$. Indeed, every function f such that $f(0^n) = 1$, $f(10^{n-1}) = f(010^{n-2}) = \cdots = f(0^{n-1}1) = 0$ is fully sensitive because $s(f, 0^n) = n$; however, we can assign $f(\mathbf{x}) = 0$ to as many of the unused $2^n - n - 1$ inputs $\mathbf{x}$, which shows that full sensitivity is "local". The minimal full sensitivity is achieved when $f(0^n) = 1$ and $f(\mathbf{x}) = 0$, for $\mathbf{x} \neq 0^n$. The same argument works for every $\mathbf{x} \in \{0, 1\}^n$ and 0, 1 instead of 0^n and 1.

The properties of "sensitivity" necessary in what follows require some "uniformity" for all points concerning their "neighbourhoods", not just one "isolated" sensitive point, as in the case of complete sensitivity. To address this issue, we define two (stronger) forms of sensitivity satisfying this requirement in the following.

A function f of $n > 1$ variables is

(a) sensitive if it is not constant 0 and for all $\mathbf{x}, \mathbf{y} \in \{0, 1\}^n$ with $d(\mathbf{x}, \mathbf{y}) = 1$ and $f(\mathbf{x}) = 1$, we have $f(\mathbf{y}) = 0$.

(b) strongly sensitive if for all $\mathbf{x}, \mathbf{y} \in \{0, 1\}^n$ with $d(\mathbf{x}, \mathbf{y}) = 1$, we have $f(\mathbf{x}) = \overline{f(\mathbf{y})}$.

It is clear that strong sensitivity implies sensitivity, but the converse

[100] A sensitivity measure has been studied as the complexity of Boolean functions [239] and is the subject of the Sensitivity Theorem [230].

implication is false. Indeed, the function $f(00) = 1, f(\mathbf{x}) = 0$, for $\mathbf{x} \neq 00$ is sensitive as $d(00, 01) = d(00, 10) = 1, f(00) = 1 \neq f(01) = f(10)$, but not strongly sensitive as $f(01) = f(11) = 0, d(01, 11) = 1$. Sensitivity implies full sensitivity because by hypothesis, there exists $\mathbf{x}$ such that $f(\mathbf{x}) = 1$, so for every $\mathbf{y} \neq \mathbf{x}$ with $d(\mathbf{x}, \mathbf{y}) = 1$, we have $f(\mathbf{y}) = 0$ showing that f is fully sensitive. Finally, R_2^1 is fully sensitive but not sensitive because $R_2^1(01) = R_2^1(11) = 1, d(01, 11) = 1$.

A more general result is true: for every f function with $n > 1$ variables, the following two conditions are equivalent:

(a) The function f is sensitive.
(b) For every distinct $\mathbf{x}, \mathbf{y} \in T(f)$, $d(\mathbf{x}, \mathbf{y}) \geq 2$.

For the direct implication, we assume by absurdity the existence of $\mathbf{x}, \mathbf{y} \in \{0, 1\}^n$ with $d(\mathbf{x}, \mathbf{y}) = 1$ and $f(\mathbf{x}) = f(\mathbf{y}) = 1$. Then, by (a) $f(\mathbf{x}) = \overline{f(\mathbf{y})}$, a contradiction. Conversely, if we assume by absurdity that there exist $\mathbf{x} \neq \mathbf{y}, \mathbf{x}, \mathbf{y} \in T(f)$ such that $d(\mathbf{x}, \mathbf{y}) = 1$, then by (b), $d(\mathbf{x}, \mathbf{y}) \geq 2$, a contradiction.

In particular, no (strongly) sensitive function is computed by a single-layer PNN. Indeed, consider a sensitive function f of $n > 1$ variables. Let us take $\mathbf{x}', \mathbf{x}'' \in T(f)$, hence by the equivalence above, $d(\mathbf{x}', \mathbf{x}'') \geq 2$. Then, there exist $1 \leq i < j \leq n$, $\mathbf{u}, \mathbf{v}, \mathbf{z}$ such that $\mathbf{x}' = \mathbf{u}x_i\mathbf{v}x_j\mathbf{z}$, $\mathbf{x}'' = \mathbf{u}\bar{x}_i\mathbf{v}\bar{x}_j\mathbf{z}$. We now choose $\mathbf{y}' = \mathbf{u}\bar{x}_i\mathbf{v}x_j\mathbf{z}$, $\mathbf{y}'' = \mathbf{u}x_i\mathbf{v}\bar{x}_j\mathbf{z}$. We note that $d(\mathbf{x}', \mathbf{y}') = d(\mathbf{x}'', \mathbf{y}'') = 1$ and $\mathbf{x}' + \mathbf{x}'' = \mathbf{y}' + \mathbf{y}''$, $\mathbf{x}', \mathbf{x}'' \in T(f)$, $\mathbf{y}', \mathbf{y}'' \in F(f)$. The conclusion follows from [23, Theorem 3.7].

Multi-layer PNNs have more computational power than single-layer PNNs. A multi-layer PNN consists of an input layer, intermediate (hidden) layers and an output layer [246]; see Figure 4.14. Figure 4.15 presents a multi-layer PNN with a single hidden layer.

A multi-layer PNN $P : \mathbf{R}^{n_1} \to \mathbf{R}^{n_L}$ is defined by $L-1$ mappings acting on a sequence of spaces $(\mathbf{R}^{n_1}, \mathbf{R}^{n_2}, \ldots, \mathbf{R}^{n_L})$, $P^1 : \mathbf{R}^{n_1} \to \mathbf{R}^{n_2}, P^2 : \mathbf{R}^{n_2} \to \mathbf{R}^{n_3}, \ldots, P^{L-1} : \mathbf{R}^{n_{L-1}} \to \mathbf{R}^{n_L}$, where each P^i $(1 \leq i < L)$ consists of j PNNs defined by:

$$P^{i,j}_{\theta^i_j, \mathbf{w}^i_j}(\mathbf{a}^{i-1}) = \begin{cases} 1, \; \textit{if} \; \sum_{k=1}^{n_i} w^i_{jk} a^i_k \geq \theta^i_j, \\ 0, \; \textit{otherwise}, \end{cases}$$

such that a) $\mathbf{a}^0 = \mathbf{x} \in \mathbf{R}^{n_1}$, $\mathbf{a}^1 \in \mathbf{R}^{n_2}, \ldots, \mathbf{a}^{L-1} \in \mathbf{R}^{n_L}$, b) $\mathbf{w}^i \in \mathbf{R}^{(n_{i+1} \times n_i)}$ denotes the weight matrix connecting i^{th} layer to $(i+1)^{th}$ layer; $\mathbf{w}^i_j$ is the jth row of matrix $\mathbf{w}^i$, c) $\theta^i \in \mathbf{R}^{(n_{i+1} \times 1)}$ is the threshold vector and θ^i_j is the jth component of this vector, d) n_i is the number of units in the i-th layer ($1 \leq i < L$) and n_L is the number of units in the output layer. For more details, see [316].

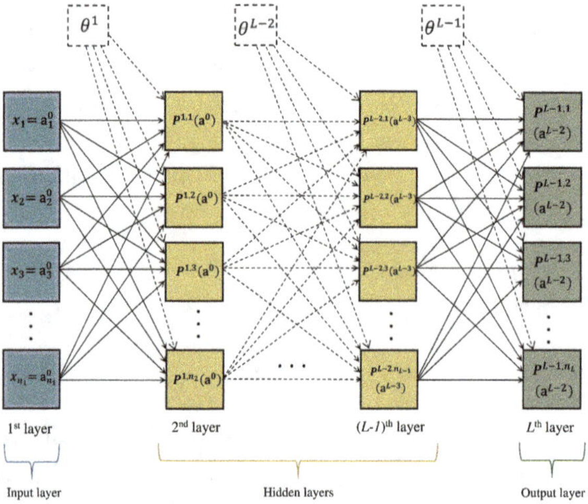

Fig. 4.14: Multi-layer PNN with an input layer, $L-2$ hidden layers and an output layer

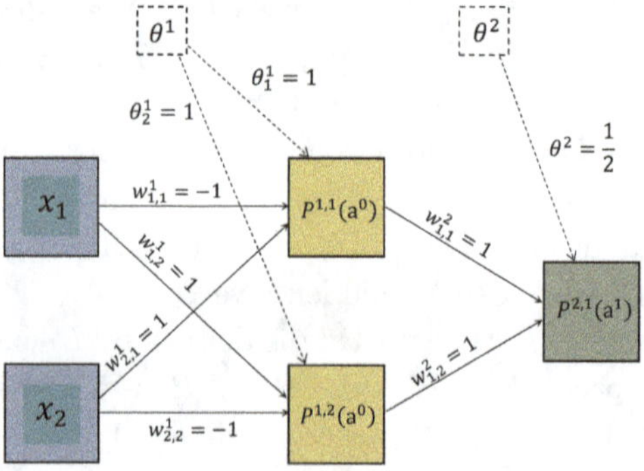

Fig. 4.15: A multi-layer PNN with a single hidden layer computing $PARITY_2$

The mappings above feed the input patterns into the hidden layers to categorise different classes in the output layer. When we have just one unit in the output layer, the multi-layer PNN is called a binary classifier.

Theorem. *Every sensitive function of $n > 1$ variables is computed by a multi-layer* PNN *with a single hidden layer and $\#(T(f))$ hidden units.*

Recall that a DNF formula consists of a disjunction of conjunctions:

$$\bigvee_{1 \leq i \leq k} \left(x_1^i \wedge x_2^i \cdots \wedge x_n^i\right), \text{ and } x_j^i = x_j \text{ if } x_j^i = 1; \text{ otherwise } x_j^i = \bar{x}_j,$$

where $f(x_1^i, x_2^i, \ldots, x_n^i) = 1$ for $1 \leq i \leq k$. Fix $k = \#(T(f))$. Based on the DNF formula, f can be computed by a multi-layer PNN with a single hidden layer where each $(x_1^i \wedge x_2^i \cdots \wedge x_n^i)$ is mapped to a unit in the hidden layer for computation [378, p. 3]. Therefore, by Theorem 7.1 [23, p. 74-83][101], a multi-layer PNN with a single hidden layer can compute any function with k hidden units.

By sensitivity and the equivalence between the conditions (a) and (b) above, for all $\mathbf{x}, \mathbf{y} \in T(f)$, we have $d(\mathbf{x}, \mathbf{y}) \geq 2$, so the points in $T(f)$ cannot be combined to reduce the DNF formula [319, p. 162]; hence the number of hidden units cannot be decreased. As each true point is a unit in the hidden layer, the hidden layer cannot have less than k units; see [378].

As a consequence of the above theorem, we deduce that every function $PARITY_n$ with $n > 1$ is computed by a multi-layer PNN with a single hidden layer and exactly 2^{n-1} hidden units [94]. In [275], a similar result was proven for a more complicated function: n variables cannot be multiplied using fewer than 2^n perceptrons in a multi-layer PNN with a single hidden layer.

In Figure 4.15, we present a multi-layer PNN with a single hidden layer and exactly $2^{2-1} = 2$ hidden units that computes $PARITY_2$.

Theorem. *For every $n > 2$ there exist $2^{2^{n-2}} - 2$ sensitive, not strongly sensitive functions which are computed by multi-layer* PNN*s with a single hidden layer and an exponential number of hidden units.*

For justification, consider a strongly sensitive function f (for which $\#T(f) = 2^{n-1}$) and remove from $T(f)$ a subset of P containing 2^{n-2} points. Then for every non-empty subset $S \subset P$ consider the function f_S whose set of true points is $T(f_S) = (T(f) \setminus P) \cup S$. Given the above results, every function f_S is sensitive, not strongly sensitive, and as $\#(T(f_S)) > 2^{n-2}$,

[101] Every function of $n > 1$ variables can be computed by a multi-layer PNN with a single hidden layer.

every multi-layer PNN with a single hidden layer that computes it has an exponential number of hidden units. Furthermore, the number of all functions f_S is at least $2^{2^{n-2}} - 2$.

How "large" is the set of sensitive functions of $n > 2$ variables computed by multi-layer PNNs with a single hidden layer and an exponential number of hidden units?

> **Theorem.** *The set of sensitive functions of $n > 2$ variables computed by multi-layer* PNN*s with a single hidden layer and an exponential number of hidden units is exponentially larger than the set of threshold functions.*

Indeed, the set of threshold functions of $n > 1$ variables has less than 2^{n^2} functions [23, Theorem 4.3], while by the previous Theorem, the set of functions which are computed by multi-layer PNNs with a single hidden layer and an exponential number of hidden units has at least $2^{2^{n-2}} - 2$ functions.

The next natural question is: Which functions are strongly sensitive? To answer this question, we first note the following invariants of strongly sensitive functions: If f is a strongly sensitive function of $n > 1$ variables, then the functions $\bar{f}$, f_π (where π is a permutation of the set $\{1, 2, \ldots, n\}$) defined by $\bar{f}(\mathbf{x}) = \overline{f(\mathbf{x})}$, $f_\pi(x_1 x_2 \ldots, x_n) = f(x_{\pi(1)} x_{\pi(2)} \cdots x_{\pi(n)})$ are also strongly sensitive.

Consider now a function f of $n > 1$ variables. The following statements are equivalent:

(a) f is strongly sensitive.
(b) The function $g_f(\mathbf{x}, x_{n+1}) = XNOR(f(\mathbf{x}), x_{n+1})$ is strongly sensitive.
(c) The function $h_f(\mathbf{x}, x_{n+1}) = XOR(f(\mathbf{x}), x_{n+1})$ is strongly sensitive.

Assume first that (a) is true and take $\mathbf{x}x_{n+1}, \mathbf{y}y_{n+1} \in \{0,1\}^{n+1}$ such that $d(\mathbf{x}x_{n+1}, \mathbf{y}y_{n+1}) = 1$. Permuting the variables and using the above invariants, we can assume that $x_{n+1} = y_{n+1}$ and $d(\mathbf{x}, \mathbf{y}) = 1$. From the sensitivity of f it follows that $f(\mathbf{x}) = \overline{f(\mathbf{y})}$, hence we have:

$$g_f(\mathbf{x}, x_{n+1}) = f(\mathbf{x})\, x_{n+1} \vee \overline{f(\mathbf{x})}\, \overline{x_{n+1}} = \overline{f(\mathbf{y})}\, x_{n+1} \vee f(\mathbf{y})\, \overline{x_{n+1}} = \overline{g_f(\mathbf{y}, x_{n+1})},$$

so g_f is strongly sensitive.

Next assume that (b) is true and take $\mathbf{x}, \mathbf{y} \in \{0,1\}^n$ such that $d(\mathbf{x}, \mathbf{y}) = 1$. By (b) $g_f(\mathbf{x}, x_{n+1}) = \overline{g_f(\mathbf{y}, x_{n+1})}$ because $d(\mathbf{x}x_{n+1}, \mathbf{y}x_{n+1}) = 1$. Indeed, if by absurdity $f(\mathbf{x}) = f(\mathbf{y})$, then $g_f(\mathbf{x}, x_{n+1}) = f(\mathbf{x})\, x_{n+1} \vee \overline{f(\mathbf{x})}\, \overline{x_{n+1}} = f(\mathbf{y})\, x_{n+1} \vee \overline{f(\mathbf{y})}\, \overline{x_{n+1}} = g_f(\mathbf{y}, x_{n+1}) \neq \overline{g_f(\mathbf{x}, x_{n+1})}$, a contradiction.

Finally, using again the invariants above, g_f is strongly sensitive if and only if $h_f = \overline{g_f}$ is strongly sensitive, that is, (b) is equivalent to (c).

The following relations are true for all $x, y, z \in \{0,1\}$:

(1) $XOR(x, XOR(y, z)) = \overline{XOR}(x, \overline{XOR}(y, z))$,
(2) $XOR(x, \overline{XOR}(y, z)) = \overline{XOR}(x, XOR(y, z))$.

Using them, we can prove that the functions $PARITY_n$ and $\overline{PARITY_n}, n > 1$ *are the only strongly sensitive functions.* Clearly, $PARITY_n$ and and $\overline{PARITY_n}$ are strongly sensitive. If f_n is a strongly sensitive function of $n > 2$ variables, then $f_n(x_1, \ldots, x_{n-1}, x_n) = XOR(f_{n-1}, x_n)$, where $f_{n-1}(x_1, \ldots, x_{n-1}) = f_n(x_1, \ldots, x_{n-1}, 0)$. Hence, f_{n-1} is also strongly sensitive. In this way, we get the sequence of strongly sensitive functions $f_{n-1}, f_{n-2}, \ldots, f_2$ satisfying the relations

$$f_i(x_1, \ldots, x_{i-1}, x_i) = XOR(f_{i-1}, x_i). \tag{4.9}$$

Out of all 16 functions of 2 variables, only two, $PARITY_2, \overline{PARITY_2}$, are strongly sensitive. Going backwards via (4.9) and using the equivalences (a) – (c), we infer that every strongly sensitive function of $n > 2$ variables can be obtained by $n - 1$ compositions of XOR and $\overline{XOR}$. Invoking the invariances, we deduce that in the set of 2^{n-1} functions obtained from all compositions of XOR and $\overline{XOR}$, there are only two distinct functions, $PARITY_n$ and $\overline{PARITY_n}$. It follows that *every strongly sensitive function* f *of* $n > 1$ *variables has* $\#(F(f)) = 2^{n-1}$.

No threshold function is sensitive, that is, "there exists $\mathbf{x}, \mathbf{y} \in \{0,1\}^n$ such that $d(\mathbf{x}, \mathbf{y}) = 1$ we have $f(\mathbf{x}) = f(\mathbf{y})$"; this property seems to be a weak form of robustness. The condition "for every $\mathbf{x}, \mathbf{y} \in \{0,1\}^n$ such that $d(\mathbf{x}, \mathbf{y}) = 1$ we have $f(\mathbf{x}) = f(\mathbf{y})$" is too strong, as it is satisfied only by the constant functions. A better definition is: the function f is *robust* if for every $\mathbf{x} \in \{0,1\}^n$ there exists $\mathbf{y} \in \{0,1\}^n$ such that $d(\mathbf{x}, \mathbf{y}) = 1$ and $f(\mathbf{x}) = f(\mathbf{y})$. For example, every function f with $T(f) = \{\mathbf{x}, \mathbf{y}\}$ and $d(\mathbf{x}, \mathbf{y}) = 1$ is robust and threshold. Furthermore, the functions R_n^1 and R_n^2 are robust and threshold. Monotone functions – i.e. those functions which for every $\mathbf{x} \leq \mathbf{y}$ (that is, for every $1 \leq i \leq n$, $x_i \leq y_i$) we have $f(\mathbf{x}) \leq f(\mathbf{y})$ – are another class of interesting functions. Monotone non-constant functions are robust, but not all are computable by single-layer PNNs. The following cardinality argument proves it: the set of threshold functions of $n > 1$ variables has less than 2^{n^2} functions [23, Theorem 4.3], which is a smaller subset of the set of monotone functions whose cardinality is the Dedekind number $D_n \geq 2^{\binom{n}{\lfloor n/2 \rfloor}}$ [379], hence the result.

A remarkable limit of machine learning was discovered in [39]: the authors proved that learnability could not be decided in **ZFC**, the standard formal system for mathematics. The reason for this undecidability is based on the profound works of K. Gödel [196] and P. Cohen [131], which is different from incomputability.

The Continuum Hypothesis is the first problem on Hilbert's famous list presented at the International Congress of Mathematics in Paris in 1900, also referred to in Chapter 2. The problem is simple: how many points on a line are there? This seemingly simple question turns out to be deeply intertwined with most of the interesting open problems in set theory, a field of mathematics aiming to be its foundation. Most objects in mathematics are infinite, and set theory is indeed just a theory of the infinite. There are infinitely many natural numbers $0, 1, 2, 3\ldots$, lines, squares, circles in the plane, polyhedra in the space, strings, algorithms, etc. But not all infinities are equal; there are different degrees of infinity. The sets of strings and algorithms are countable because they can be "paired" bijectively with the set of natural numbers; the set of real numbers – also called the "continuum" – is not countable, as Cantor proved in the early 1870s. The big question is: are there any infinities between the two infinities of natural numbers and real numbers? According to the Continuum Hypothesis, *there is no third infinity between the two.* Gödel, like Hilbert, believed that the Continuum Hypothesis would be solved, hence his Incompleteness Theorem, though showing the existence of undecidable statements, has nothing to do with it. In 1937 Gödel showed that the Continuum Hypothesis cannot be proved false in **ZFC**. In a letter to Gödel from 24 April 1963, Cohen communicated the converse result: we cannot prove that the Continuum Hypothesis is true in **ZFC**. Hence, the Continuum Hypothesis cannot be proven right or wrong in **ZFC**; it is independent of **ZFC**.

What do these esoteric questions have to do with machine learning? Apparently nothing. The answer is "quite a lot". AI algorithms use mathematics and statistics to enable machines to compute without explicit programming, also known as "hard coding". Pattern recognition, regression and clustering are problems which machine learning can quickly solve with the following method:

> Approximate a target concept given a bounded amount of data about it.

The article [39] studies a learning problem called "estimating the maximum" (EMX), and here is an example. Consider a website used by a set of users denoted by X. The owners of the website post on it ads chosen from a given pool. Each ad A in the pool targets a particular population of users $FA \subseteq X$. For example, if A is a song ad, then FA is the collection of the song fans. The goal is to place an ad whose target population visits the site most frequently without knowing in advance which visitors will visit the site.

Is this problem solvable by AI? The authors of the paper proved that in certain cases, an AI solution to the EMX problem is equivalent to a solution to the Continuum Hypothesis, which, by the work of K. Gödel and P. Cohen, is *unprovable*. This theoretical result, which may not be (yet?) related to a practical application, demonstrates a limit of learnability. One may ask: what is the reason that learnability can neither be proven nor refuted? The authors' answer is:

> A closer look reveals that the source of the problem is in defining learnability as the existence of a learning function rather than the existence of a learning algorithm. In contrast with the existence of algorithms, the existence of functions over infinite domains is a (logically) subtle issue.

Another limit of AI is related to "stability".[102] The article [132] proved that the current deep learning methods suffer from instability, even when universal approximation properties guarantee the existence of stable NNs. The authors proved that for some well-conditioned problems in scientific computing, *one could show the existence of NNs with great approximation qualities despite the fact that there does not exist any algorithm, even randomised, that can train (or compute) such an NN.*

These results lead to a classification theory describing conditions under which an algorithm can compute (stable) NNs with a given accuracy. They also reveal another aspect of AI: its difficulty, sometimes impossibility, in recognising when things are wrong. Sometimes, it's even more difficult for an AI system to realise when it's making a mistake than to produce a correct result.

[102]Other results have been reported for underfitting [352] and superintelligence [17].

4.15 Free will and randomness

People seem to behave, consciously or not, as if they possess free will; they don't feel like automata, and most certainly, they don't treat one another as they might treat robots or programs.

Free will is difficult to understand or explain, and the literature on free will is huge; still, frequently, free will is ignored, as, for example, in the eight-volume *Encyclopedia of Psychology* edited by A. E. Kazdin [30, p. 3].

In this section, we discuss only two aspects of free will: its relations to randomness and physics.

Many authors claim that randomness is incompatible with freedom as an obvious fact [165]. Here is an argument:

> Yet if an uncaused action is a random happening, then this no more comports with human value than does determinism. Random acts and caused acts alike seem to leave us not as the valuable originators of action but as an arena, a place where things happen, whether through earlier causes or spontaneously. Clearly if our actions were random, like the time of radioactive decay of uranium 238 emitting an alpha particle, their being thus undetermined would be insufficient to ground human value or provide a basis for responsibility and punishment. [310, pp. 91–92]

The allegedly intimate relation between free will and (absence of) randomness has been examined from the very beginning of the debate about free will. Despite its important role in this debate, the notion of randomness itself is often left unexplicated or identified with chance or even indeterminism [165]. In what follows, we will use the product conception of randomness [169]: random happenings are haphazard and lack any discernible or explicable pattern; see Section 2.5. This notion of randomness applies to individual objects and is of most theoretical and philosophical interest in understanding the relation between free will, indeterminism, and randomness. Following [96], we will argue that randomness in an agent's choices is *independent* of whether she is free to make those choices; hence *free will is compatible with randomness.*

In the vast literature on free will, a lot of misunderstanding comes from the fact that free will has been understood in various ways. In the following argument, we will propose a simple, two-stage, contextual (not absolute)

Valerian definition[103] of free will. This definition models the idea that *an agent has free will in some context if it can make a decision that is not completely determined by or the result of previous events.* More precisely, an agent A acted freely in the context $\mathfrak{C}$ with respect to the set of objects $O_{\mathfrak{C}}(A)$ if A could have acted differently to the way it did and A had full control over the outcome of the choice. This means that in the context $\mathfrak{C}$:

- **(P)** The set of objects $O_{\mathfrak{C}}(A)$ available to A contains at least two elements and every choice for A is possible in the given context.
- **(C)** The agent A has full control over which of the objects in $O_{\mathfrak{C}}(A)$ to choose.

The Possibility Assumption, **(P)**, states that in the context $\mathfrak{C}$ there are at least two objects in $O_{\mathfrak{C}}(A)$ available to A to choose from; hence the choice is possible, meaningful and can lead to different outcomes. Furthermore, $O_{\mathfrak{C}}(A)$ depends on the context $\mathfrak{C}$ and can vary with $\mathfrak{C}$ – by changing the context $\mathfrak{C}$, some elements may disappear, and new elements can be added to $O_{\mathfrak{C}}(A)$. The identity of $O_{\mathfrak{C}}(A)$ may not be entirely known by A; hence, A may not exhaustively know what, in the context, is possible for her. The Choice Assumption, **(C)**, posits that the agent has complete control over choosing an option. The idea of full control may seem contentious since it suggests i) that such an agent has powers that no ordinary agent has (for example, the power to generate and arrange at will the mental states involved in evaluating the various options faced by the agent) and ii) that the options open to the agents must be fully transparent to her, something we denied. We say nothing about i) here, apart from to note that 'full' is rather like 'perfectly' in 'perfectly flat'. For ii), we require only that the agent is aware of at least two members of $O_{\mathfrak{C}}(A)$, and that once she is aware of a member of $O_{\mathfrak{C}}(A)$, she has the intentional ability to select one. According to **(C)**, in the case of free choice, the contextual possibility of an option is an intentional contextual possibility, and so it involves intentional ability in Mele's sense [295]. These two assumptions satisfy a broadly libertarian view of free will, i.e. to be free, an agent must be able to do otherwise from what she did.

The contextual availability of every element in $O_{\mathfrak{C}}(A)$ is essential to free choice. For example, if A is a human being, then the set $O_{\mathfrak{C}}(A) =$ {to be immortal, not to be immortal} violates the condition of contextual

[103] Dennett [157, p. 293] called his model of free will "Valerian" after P. Valéry who was quoted as saying that it takes two to invent anything. The one makes up combinations; the other one chooses.

possibility: "to be immortal" is not available to A if A is human (although it may be available to certain mythical heroes). So A cannot act freely in such a context $\mathfrak{C}$.

Is the above definition incompatible with the "mainstream" literature on free will? The definition is contextual, not absolute nor global. It agrees with the conceptions of free will advanced by many other authors, including the quantum physicists' notion of free will [422], and it does not assume any strong forms of rationality or moral responsibility. Finally, it satisfies the conditions of van Inwagen's classic (if contentious) argument for incompatibilism [402, p. 70] and is supported by most of Dennett's [157] (Ch. 15) reasons in favour of his two-stage model.

In what follows, "free will" will always refer to the above definition; if the context $\mathfrak{C}$ is understood, we may write $O(A)$ instead of $O_{\mathfrak{C}}(A)$.

The next question is: Is free will compatible with determinism? That is, can free will exist in a deterministic universe? While this is a crucial point of debate among those writing on free will, it is not central to our discussion. Broadly, determinism can be taken to be the view that given the state of the universe at time t, the way things go thereafter is fixed as a matter of natural law [225]. There are different variations, but we will not analyse these claims here except to point out that supporters of such forms of compatibilism are left with the task of explaining how **(P)** holds in a universe in which the event of an agent's making a particular choice is fixed, for every earlier time t, by the state of the universe at t, including times t that preceded the existence of the agent.

If an indeterministic choice is random, how can we still say an agent has free will? The notion of chance, sometimes described as 'inexplicable' and 'unpredictable', is also related to randomness; but in which way? According to [280],

> ... randomness ... is in the interface between our theoretical descriptions and 'reality' as accessed by measurement. Randomness is unpredictability with respect to the intended theory and measurement.

Is indeterminism necessary for randomness? The answer is negative. Coin-tossing and software-generated randomness are (weak) forms of randomness which are algorithmically produced – no form of indeterminism is required. Both algorithmic information theory [61] and the practice of generating randomness show that randomness (of any form) is: a) defined by avoiding some algorithmically defined classes of patterns and b) not produced algorithmically. Here are two examples, one mathematical and one physical. The infinite sequence given by the binary expansion of Ω_U discussed in Section 2.8, is uniquely determined by U, but this sequence is "highly random". Randomness appears when we "don't know" that the sequence is the binary expansion of the Omega number of U. This phenomenon is general because, in every sequence, there are patterns and correlations. In the same way, in quantum mechanics, the Schrödinger equation describes how the quantum state of some physical system changes with time: the wave function evolution is deterministic. Quantum randomness appears when we observe/measure certain individual quantum observables, for example, those that are *value indefinite* [14]. There is a loose analogy between the above two examples. The description of Ω_U by U corresponds to the Schrödinger equation. A first analogy appears when we compare the processes of computing the bits of Ω_U and measuring value indefinite observables in a quantum experiment E precisely described in [14]. One can prove that the sequence of bits of Ω_U cannot be algorithmically computed (given U) and, similarly, the sequence of bits obtained by performing ad infinitum the quantum experiment E cannot be algorithmically computed (given full information about the experiment [14]). A second, more interesting analogy appears when we compare the unpredictability of individual bits. An individual bit of Omega is maximally Martin-Löf unpredictable; a bit obtained in the quantum experiment E is maximally unpredictable in the sense of [14]. The computer U cannot give the exact value of any bit of Ω_U, and similarly, the Schrödinger equation cannot give the exact result of an individual measurement.

Conversely, is randomness necessary for indeterminism? Again the answer is negative: there are cellular automata and a plethora of computing machines that work in non-deterministic ways without using randomness. Quantum computing is a paradigm which works probabilistically without any randomness. The same is true if indeterminism is considered causal: nothing can be inferred about what kind of regularity (or absence of regularity) the outcomes may have given only the lack of causal determination in producing them.

Having seen that true randomness does not exist even mathematically, see Section 4.10, let alone in nature, we can now examine the compatibility between randomness and freedom more closely.

First, all arguments against the compatibility between free will and randomness based on "pure chance", "pure randomness", and "true randomness" are unsound because they rest on a vacuous concept. An example is the argument based on Hume's and Schlick's ontological thesis, according to which there is nothing intermediate between chance and determinism. In Eddington's words ([170, p. 182]):

> There is no half-way house between random and correlated behavior. Either the behavior is wholly a matter of chance, in which case the precise behavior within the Heisenberg limits of uncertainty depends on chance and not volition. Or it is not wholly a matter of chance, in which case the Heisenberg limits ... are irrelevant.

Popper [322, p. 227] disagreed:

> Hume's and Schlick's ontological thesis ... seems to me not only highly dogmatic (not to say doctrinaire) but clearly absurd.

The reference to "correlated behavior" shows that Eddington was working with the product notion of randomness (or perhaps an amalgam of the process and product notions). And as we have seen, randomness, in that sense, only permits degrees of randomness. Our choices and actions are never the product of "pure randomness".

A more interesting but still unsound argument for the incompatibility between free will and randomness is the following: randomness exists (in various degrees), so if an agent's actions are caused by randomness, the agent lacks control, so the assumption **(C)** is violated. In detail, the argument runs as follows.

(1) Assume **(P)**.
(2) An agent A has free will with respect to $O(A)$ in the context $\mathfrak{C}$ if the assumption **(C)** is satisfied.
(3) So A has ultimate control of which object in $O(A)$ to choose.
(4) If the object was chosen randomly (to some degree), then no one had full control of which object in $O(A)$ was chosen.
(5) Hence, A cannot have full control on which object in $O(A)$ to choose.
(6) Therefore, A has no free will with respect to $O(A)$ in the context $\mathfrak{C}$.

The argument may have some force if one could explain how the selected object was chosen sufficiently randomly to prevent A from having control over its decision. Unfortunately, this explanation is not given. Instead, it is often claimed that **(P)** is inherently able to provide this [20, 382].

As we already noted, however, randomness does not "float around" and is not something that is somehow "imposed on the agent". Randomness is just produced and then used by the agent in coming to a decision. To make random decisions, the agent needs to use a random generator, which produces random bits of a certain quality (but never "truly random bits"); there is no alternative.

According to the adopted definition of free will, the detailed process used by the agent A to choose an object in some context $\mathfrak{C}$, where more than one object is available for selection, is irrelevant: with **(P)** in place, all that is needed for her to choose freely is the satisfaction of condition **(C)**. To make a decision, using a random generator is no different than using the advice of a friend or getting more information from Wikipedia!

We will look closely at the random selection process for a better understanding. We assume that **(P)** is satisfied and consider how a random generator may interact with an agent's decision-making process and how the quality of the randomness generated may affect the agent's freedom. Regarding the former, we discuss four possible interactions between the agent and the random generator.

Suppose that in a context $\mathfrak{C}$, $O(A) = \{0, 1\}$, A chooses a random generator G with two possible outputs, 0 or 1, and then operates as follows:

(0) A uses G which outputs x, but ignores the output and picks an element of $O(A)$.
(1) A uses G which outputs x, and, after first deciding whether to use x to pick an element of $O(A)$, A then picks an element of $O(A)$.
(2) A uses G which outputs x and continues as follows:
   - A uses x to determine whether to use G or not,
   - depending on x, A makes no use of G or uses G to produce another (independent) output y which becomes its decision (in the last case G actually takes the decision on A's behalf).
(3) A uses G which outputs x and its decision is x (G is used to take the decision on A's behalf).

In case (0), it is clear that A's freedom is not hindered by randomness. Still, it is worth pointing out that there was a random element in her decision process (though it did not impact A's decision). To show that case (1) does not undermine A's freedom, we restate that the information that A uses to make a decision is irrelevant to A's decision being free: A merely asks G for advice. Cases (0) and (1) are identical if A picks something other than G's output. In either of these cases, however, A can consistently choose any element in $O(A)$ and, regardless of what G outputs, A has the final say on the decision. Hence, neither of these cases will disturb A's freedom.

Cases (2) and (3) are more challenging to analyse. Case (2) is a hybrid between the earlier cases and the substantially more severe case (3). In (2), A operates G to generate a random bit. Depending on what this generated bit was, G either stops (leaving A to make the decision) or uses G to generate another random element of $O(A)$ and chooses this element on A's behalf. Thus (2) will either reduce to (0) or (3) depending on the result of the first computation. The point is that in cases (2) and (3), once A starts G, the chosen object may be decided by G rather than A. Is **(C)** fulfilled in these cases? Does the quality of random bits matter (e.g. if they form an incomputable sequence)?

These cases seem to violate assumption **(C)** because the agent gives up its final decision, not because the agent gives up its final decision to a random process. Asking another agent to decide on its behalf is no different than asking a random generator. Notice that whether A retains its freedom in asking another agent B to make its decision is a delicate issue, which is, in practice, judged on a case-by-case basis. This shows that from the point of view of free choice, the role of B is as detrimental to A's freedom as the role of G. Giving up the freedom to randomness is as harmful as giving up the freedom to any other agent: A retains its freedom when it gives up its decision to B if and only if A retains its freedom when it gives up its decision to G.

Similarly, even if A does not voluntarily give up its decision, another agent B choosing an element of $O(A)$ is no different than some random generator G doing the same. Thus, the fact that indeterminism allows for randomness should not lead us to conclude that free will is impossible any more than the fact that other free agents can choose on behalf of others.

In short, the arguments above showed that indeterminism is a necessary condition for free will and that, despite the threat from the notion of randomness, it is possible for agents to be free in some contexts under indeterminism. Randomness is compatible with free will as long as it is metaphysically possible. The arguments are relative and do not answer the main philosophical question about free will, namely, does free will exist and, indeed, can it exist? This result is similar to The Free Will Theorem, which roughly says that under three specific assumptions, "If experimenters have free will, then so do elementary particles." [134]. Indeed, "Although the Free Will Theorem can't prove if we have free will, it does have a fundamental consequence: if the universe is deterministic, and a function of the past always describes a particle's behaviour, then we can't have free will" [394].

The compatibility between randomness and free will could have consequences for psychiatry [238]:

> ... if indeterminism opens up the capacity of doing otherwise and engaging in free actions, we propose that mental illnesses, where the capacity to do otherwise is often curtailed (e.g. under the effect of delusion, substance dependence, obsession, depressive cognitive distortions), may be associated with overdeterminism that may originate from biological and/or psychological processes. In this sense, some mental disorders may be conceived as pathologies of freedom.

4.16 Is there a universal image generator?

Synthetic pattern generation procedures have various applications, and several approaches (fractals, L-systems, etc.) have been devised. A fundamental underlying question is: will new pattern generation algorithms continue to be invented, or is there some "universal" algorithm that can generate all (and only) the perceptually distinguishable images, or even all members of a restricted class of patterns, such as logos or letterforms?

Many algorithms can generate all possible images, but *most images are random and not perceptually distinguishable.* Counting arguments shows that the percentage of different images that such complete algorithms will generate is vanishingly tiny. Can we "fine-tune" such a complete algorithm to generate only "meaningful" images? If such an algorithm exists, it will be called *universal image generator.*

Before getting into more technical details [97], let's note that a positive answer to the question will be very important from a practical point of view. What about a negative answer? Would such an abstract result have any importance for graphics and other fields that deal with compressible signals? In essence, new representations and pattern-generation algorithms will continue to be developed even though no feasible "super algorithm" exists.

In the interesting essay *Metamagical Themas* [226], Hofstadter considered a similar but simpler question: is there an algorithm that can generate all possible letterforms?

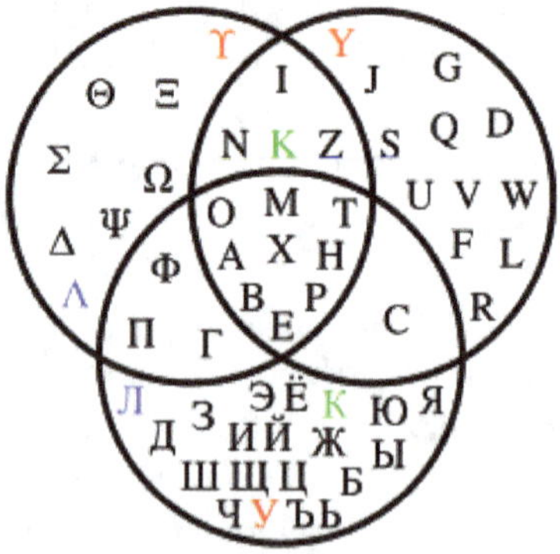

Fig. 4.16: Venn diagram of letters in the Greek, Cyrillic and Latin alphabets[104]

As it seems trivially possible to create any desired letterform with a spline drawing program such as Adobe Illustrator, the question needs to be explained. A universal generation algorithm for a class of images is an algorithm that can generate all members of that class and that never (or only very rarely) generates objects outside that class. In the case of Hofstadter's essay, one imagines a master typeface program that will generate all possible letterforms as user-specified style parameters are varied, or that generates all possible letterforms automatically with no user input. The program's outputs should be predominantly images of the intended class – a small percentage of non-letterform patterns among the outputs of a letterform generator might be acceptable.[105] However, a letterform generator that only rarely produces letterforms hardly justifies the name.

According to Hofstadter [226]

> a letter forming algorithm is *complete* if it can generate all possible letterforms, and *consistent* if it generates only letterforms and no other types of images.

Accordingly, we define a *universal algorithm* to be one that is both consistent and complete.

Hofstadter suggested that a *universal letterform algorithm does not exist.* The essay argues that a single algorithm cannot encompass the variety of unusual (but recognisable) letterforms that have been devised, as well as

[104]Source: In Wikipedia. `en.wikipedia.org/wiki/File:Venn_diagram_showing_Greek,_Latin_and_Cyrillic_letters.svg`.

[105]Challenge for the reader: why?

the infinity of letterforms that have yet to be designed, and points out that existing recognisable examples of, say the letter "A", do not even have any single feature in common [226]. Hofstadter's arguments are pretty convincing even though they do not constitute a proof.

The possibility of a universal image construction is of philosophical and practical interest. Hofstadter framed the question philosophically as an issue of whether the essence of something like a letter can be algorithmically defined. Practical graphics tasks such as designing a corporate logo could benefit from universal visual pattern generators. The existence of a universal logo generator might allow graphic designers to skip the construction of candidates and merely select from a long list of possibilities generated by the program. For another example of a practical application, consider the visualisation task of providing abstract visual icons for a large set of data. This task requires a procedure that can generate a broad range of visually distinguishable icons. Various pattern synthesis approaches (fractals, L-systems, etc.) come to mind, but choosing any one of these already restricts the range of patterns that will be generated. So, the question is:

Is there a universal image generator?

Although a program that can generate all possible letterforms, or all possible corporate logos may be conceivable, a program that generates all possible images (including, for example, a picture of you on a vacation that you will take in the future) initially seems more farfetched. As we already noted, programs that enumerate all possible images do exist, but they are inconsistent and therefore are not useful.

We start with a simpler instance of the problem. A "British Museum" algorithm can generate all possible binary strings, i.e. by an exhaustive enumeration. There are infinitely many algorithms to accomplish this, for example, the enumeration in shortlex order. In a similar way, going from 1D to 2D, all possible images can be generated simply by enumerating all pixel combinations, i.e. for an image with n pixels, do the equivalent of n nested loops, each looping over all pixel values. Infinitely many other procedures exist: for example, pick any complete image basis and enumerating all coefficient values in that basis will generate all possible images. In addition to exhaustive enumerations, the use of a "reliable" source of randomness, e.g. quantum generated randomness [14], to pick pixel values or coefficients will eventually generate any possible picture.[106]

[106]Challenge for the reader: Why?

Each of the above universal methods produces a complete set of images; are they consistent? Although, in theory, a set of pictures documenting all aspects of your life will appear in the output of these programs if you wait long enough; in fact, the images generated by these approaches will look very homogeneous – almost all images will look like the "television noise"; see Fig. 4.17.[107]

A program that, in practice, only generates noise images is of little value as a universal image generator: indeed, this is not just that most images generated with these approaches will not look like natural images, but that one would have to wait a very long time to observe anything resembling a real-world image.

Fig. 4.17: "Random television noise" images are not memorable and not easily distinguishable

It is worth characterising the magnitude of the quantities involved in this analysis. If we take a black-and-white 64^2 8-bit image as a minimal image representation, and generously assuming 8:1 lossless compression can be obtained, then the number of images is $8^{4096}/2^3 = 2^{12285}$, a huge quantity of images. Operations such as these are theoretically possible but practically impossible, are labelled *infeasible*. These computations should be distinguished from the practically achievable ones, and also from the incomputable ones, like those appearing in the Halting Problem.

A different approach is to observe that the parameterisation of image space embodied in a pixel or Fourier basis is not matched to our perception.

Most of the space in each of these representations maps to images that are indistinguishable, while the set of perceptually distinguishable images is a tiny subset of the space that random sampling, Fig. 4.16.

[107]Challenge for the reader: Use a counting argument to justify this statement.

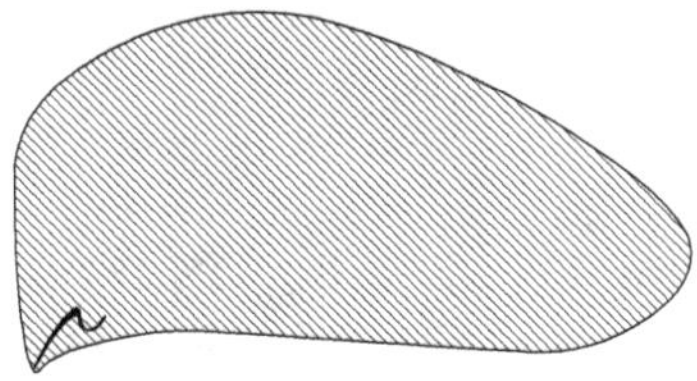

Fig. 4.18: The space of all possible images: the space of natural images (black sliver), though vast, is a tiny subset that random sampling will not locate

This suggests that consistency means that we algorithmically generate all possible *perceptually distinguishable* members of a class of images. Since our visual systems are evolved to distinguish the types of images that arise in nature we will term these perceptually distinguishable images as *natural-like* (NL) images.

Most invented images are also designed to be perceptually distinguishable and hence fall in the NL category. For example, computer graphics images of imaginary objects are usually easily distinguishable, as are many other graphic designs. On the other hand, the human visual system is not designed to distinguish pictures as that in Fig. 4.17; random-like images are not NL.

Are the complete enumerations described above consistent? Certainly not. Can they be adapted to satisfy consistency, i.e. to account for perception? In theory, this might be accomplished by defining a perceptual metric. The space would be sampled as seen through this metric, so indistinguishable areas of the image space would not be repeatedly sampled. While conceptually, we could use pairwise similarity rankings in conjunction with multidimensional scaling (or other manifold learning procedures) to identify a perceptually parameterised space, there is a stubborn underlying problem – again, the space of images is infeasibly large. Consider a set of several thousand images the subjects are to rank for similarity (a boat, a dog, a car, a book, …). This is such a small subset of the space of possible images that extrapolation to other images is not reasonable, and simultaneously, the distance between the individual images is so large that interpolation is ill-posed. Each pair of images will have significant differences in many thousands of coordinates; there is not enough information in the ranked distance to know what combination of coordinates is responsible for that distance.

Next, we consider an abstract approach that first characterises the subspace of NL images and then projects onto this subspace from a random location (image). As an approximation of this approach, take the characterisation of NL images to be piecewise smooth. An attractive continuation method, *graduated non-convexity* (GNC), has been formulated for the problem of fitting a piecewise-smooth surface to data [49]. The GNC 'weak membrane' model fits a smooth surface through the data but also allows discontinuities where the modelled surface changes by more than a threshold between pixels. In Fig. 4.19, we applied the weak membrane to an initially random (television noise) image as an approximation of this idea of projecting on the NL manifold(s). Fig. 4.19 suggests that this characterisation of "natural-like" is too simplistic and does not produce particularly distinguishable images.

These ideas can be concretely illustrated by considering JPEG images. JPEG can represent all possible distinguishable pictures, and the representation is such that perceptually insignificant information is discarded. A JPEG decoder could be adapted to generate pictures by driving it with random discrete cosine transform (DCT) coefficients. Even though JPEG discards some perceptually indistinguishable data, an attempt to employ it as an image generator will result in noise images; see Fig. 4.20.

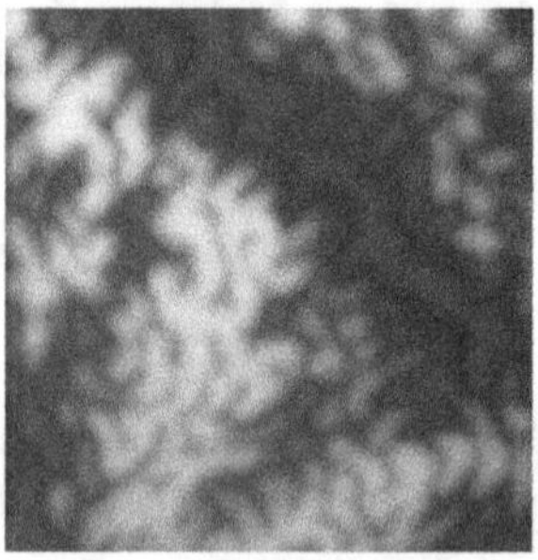

Fig. 4.19: Example of a random piecewise smooth approximation to a "natural-like" image. A collection of such images would not be particularly distinguishable

These approaches are underdeveloped because of our incomplete understanding of human perception and the character of natural images. In fact, the abstract argument indicates, *but does not prove*, that the goal of algorithmically producing all distinguishable images is not possible.

Following [97], we will use the algorithmic complexity presented in Section 2.5 to define algorithmically random strings, the main ingredient in a model of NL images.

Recall that a string x is (algorithmically) m-random if $K(x) \geq \text{length}(x) - m$, where m is a positive integer significantly smaller than $\text{length}(x)$.

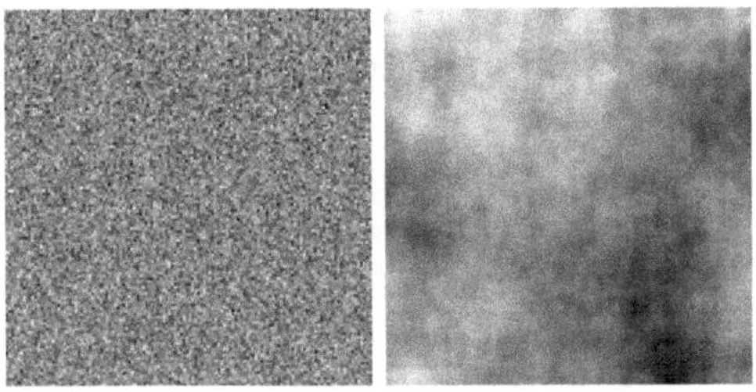

Fig. 4.20: Pseudo-random images

In Section 2.6 we proved that so there are no more than $2^{n-m}-1$ strings x of length n with $K(x) < \text{length}(x) - m$. For example, there are at most $2^{n-1} - 1$ programs of size $n - 2$ or less, so fewer than half of the strings x can have $K(x) \leq n - 1$. Likewise, less than a quarter of all strings are compressible by two or more bits, and less than $1/2^m$ of all strings have complexity less than $n-m$. This means that "exponentially few strings are exponentially compressible".

Fix two integers $k, i \geq 0$. With the exception of finitely many n, the proportion of n-bit strings x having $K(x) < k$ is smaller than n^{-i}, a quantity that (effectively) converges to 0 when n tends to infinity (Fig. 4.21). Here is a counting argument proving this assertion. Take an n such that

$$m \leq n - i \log_2 n, \tag{4.10}$$

and observe that only finitely many n do not satisfy (4.10).[108] Out of all 2^n n-bit strings, at most $2^{n-i\log_2 n}-1$ strings can have $K(x) < m \leq n-i\log_2 n$; hence for each n satisfying (4.10) the proportion of n-bit strings x having algorithmic complexity $K(x) < m$ is smaller than $2^{n-i\log_2 n}/2^n \leq n^{-i}$.

[108]Challenge for the reader: Can you justify?

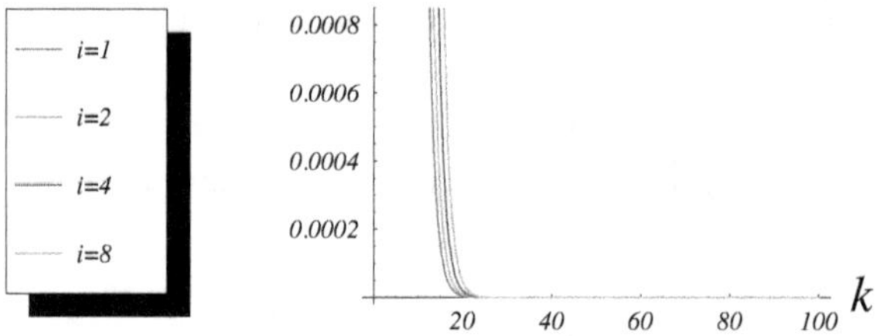

Fig. 4.21: Number of n that violate (4.10) for several i, plotted as a proportion of 2^k

The above analysis provides quantitative answers to the questions raised earlier in this section concerning the "British Museum" approach to complete image generation.

- Any complete image synthesis procedure of the "'British Museum" type, will not only generate random images such as Fig. 4.17, but these types of images will vastly outnumber the non-random ones. The reason is that almost all digital objects are incompressible.
- The fact that almost all objects are incompressible is true regardless of the basis used to represent the objects [61]. This result provides indirect proof that choosing random coefficients on some other basis (e.g. Fourier) generally produces a noise image.
- The likelihood of encountering anything other than a noise image can be evaluated. A 64^2 8-bit image is $8 \cdot 2^6 \cdot 2^6 = 32\text{k}$ bits uncompressed. An image with half of the complexity (2:1 compression) might be sufficiently compressible that it would not look like another noise image. Such an image would be smaller by 16k bits; the proportion of strings that are 16k bits less complex than a random string is $2^{-16384} \approx 8.401 \cdot 10^{-4933}$ – an infeasible minority.

We are ready to present a complexity-theoretic model of NL images. We start with the question: Are NL images compressible? The fact that the somewhat unnatural images represented in Fig. 4.17 are also not compressible suggests that the converse is also the case, i.e. that natural images are compressible. Indeed, this is certainly true – lossless compression of about 2.5:1 is commonly achieved [363], and lossy compression of 10:1 or more can be achieved with little or no perceptual degradation.

Can we conclude that most or all images that are potentially of interest are also compressible? Most images of the real world depict objects that have some formative process, and that process can be approximately or statistically modelled to obtain compression. Even some random time series that occur in nature are correlated ($1/f$ noise) or consist of intermittent events (radioactive decay) and are thus compressible, although this is not the case, in general; see [108]. Similarly, in the realm of human creations, most designs reflect some internal logic rather than being "drastically" random. Additional evidence is provided by the fact images typically have a power-law spectrum [348] and thus are compressible (it was argued that human vision takes advantage of this fact [29]).

The distinction between compressible and incompressible does not give a good criterion to define NL images. However, a "degree" of incompressibility might work. The problem is: where to have the boundary, or more precisely, how compressible are NL images?

One solution is to require that one can compress an NL image x by at least the logarithm of its length:

$$K(x) \leq \text{length}(x) - \log \text{length}(x). \tag{4.11}$$

However, NL images are "not too simple" because their complexity is not too low. For example, the images in Fig. 4.20 have low complexity. Indeed, their K complexity is about the logarithm of their length plus a constant: a pseudo-random string can be generated by a seed amplified by a short algorithm. Hence one needs a constant plus the length of the output to produce it; the output length can be encoded in $\log \text{length}(x)$ bits. This suggests the criterion: one cannot compress an NL image x by more than the logarithm of its length plus a constant:

$$K(x) > \log \text{length}(x) + \text{constant}. \tag{4.12}$$

From the above discussion, we conclude that NL images are compressible but not too simple. These two facts are expressed mathematically by (4.11) and (4.12), so we are led to formulate the following

NL Thesis. **[97]** *Every natural-like image x satisfies (4.11) and (4.12).*

Formally, the NL Thesis says that NL images can be coded by the infinite set of strings

$$\{x \mid \log \text{length}(x) + \text{constant} < K(x) \leq \text{length}(x) - \log \text{length}(x)\}. \tag{4.13}$$

Anticipating, if this NL Thesis is accepted – more arguments supporting it will be discussed later – algorithmic complexity arguments will show that that there is no single program that can generate all NL images, a result that the reader has already guessed.

Here is the mathematical argument. Fix $k \geq 0$ and consider a set of very simple images represented by the bounded algorithmic complexity strings,

$$A_k = \{x \mid K(x) \leq k\}.$$

The set A_k is computably enumerable. First, note that $K(x) \leq k$ if and only if there exists a program p with length$(p) \leq k$ and $U(p) = x$.[109] Furthermore, $U(p) = x$ if and only if there exists a time $t > 0$ such that $U(p)$ stops in time t and $U(p) = x$. We can use a method called dovetailed.[110] An enumerator generating in parallel all pairs (p, t) with length$(p) \leq k$ and $t > 0$ such that "$U(p)$ stops in time t and $U(p) = x$". Note that the above predicate depends on p, and t is computable and will enumerate all strings in A_k and only them.[111]

Furthermore, A_k is finite (so computable), but not *uniformly computable in k*. Indeed, A_k has at most $2^{k+1} - 1$ elements, so is computable. Can membership in A_k be decided by an algorithm working with parameter k (i.e. uniformly in k)? The answer is negative because the function

$$f(k, x) = \begin{cases} 1, \text{ if } K(x) \leq k, \\ 0, \text{ otherwise}, \end{cases}$$

is not computable and the reason is that the set $A = \{(x, k) \mid K(x) \leq k\}$ is not computable (for a stronger result, see [61], Corollary 5.37). This negative result holds true despite the set A_k being not only finite but very small, as shown above.

Suppose the universal image generator program exists and further, suppose that the time it requires to generate each image has some computable limit. This limit may be long, i.e. we allow the program to run for hours (or even centuries) to generate each image, but there is nevertheless some bound, perhaps one that is a function of the resolution and bit depth. Then, we shall show how to adapt this program to compute the algorithmic complexity of an arbitrary string, which is impossible.

In view of Corollary 5.37 in [61], the set of compressible strings

$$\{x \mid K(x) \leq \text{length}(x) - \log \text{length}(x)\},$$

[109] Recall that U is a universal Turing machine.

[110] In carpentry, dovetailing connects two pieces of wood by interleaving them.

[111] Challenge for the reader: Prove this claim.

is computably enumerable but not computable.[112]

This set is also very small. Indeed, a counting argument shows that the proportion of the above n-bit compressible strings in the set of n-bit strings is smaller than $2^{n-\log n}/2^n \leq 1/n$, a function which tends to 0 when n tends to infinity.

Now we can prove the main result:

> **NL Theorem.** [97] *Assume the NL Thesis. Then the set of NL images is not computably enumerable, hence no program generates all (and only all) NL images.*

Indeed, NL is coded by an infinite subset of

$$\{x \mid \log \text{length}(x) + \text{constant} < K(x) \leq \text{length}(x) - \log \text{length}(x)\},$$

so an infinite subset of the set

$$\{x \mid \log \text{length}(x) + \text{constant} < K(x)\}.$$

The last set is *immune* in virtue to Corollary 5.34 in [61]); that is, it contains no infinite computably enumerable set, hence it is not computably enumerable. In particular, NL is not computably enumerable.

How relevant is the above negative result? The answer depends on the plausibility of the NL Thesis. The specific logarithmic functions used in (4.11) and (4.12) may seem rather arbitrarily chosen. Fortunately, all the previous results hold true in a more general context. To prove this we consider the class of possible thresholds

$$T = \{f \mid f \text{ is computable}, \lim_{n\to\infty} f(n) = \infty,$$
$$\text{there exists}\, c > 0 \;(\lim_{n\to\infty} n - cf(n) = \infty)\}. \tag{4.14}$$

The threshold functions used in the NL Thesis, $f(n) = \log n + \text{constant}, g(n) = n - \log n$, are clearly in T. The functions $\log\log n, \sqrt{n}, \lfloor n/2 \rfloor$ are in T, but the function $n-2$ is not in T because it grows too fast.

The class T has a few interesting properties: a) if $f \in T$ and a is a positive integer then $af \in T$, b) if $f, g \in T$, then $\min\{f,g\}, \max\{f,g\}, f+g \in T$, c) if $f, g \in T$, then there is a constant $c > 0$ such that $g(n) < n/c - f(n)$, for almost all n. For b) note we need to prove that $f+g \in T$ provided

[112] Challenge for the reader: Can you give a direct proof?

$f, g \in T$: if $f(n), g(n)$ satisfy the second condition in (4.14) with constants c_f, c_g, respectively, then $f+g$ satisfies the condition for $c = \min\{c_f + c_g\}/2$; c) follows from b).

With a very similar argument, we can show that for all functions $f, g \in T$ with $f(n) < g(n)$ almost everywhere, the set

$$\{x \mid f(\text{length}(x)) < K(x) \leq g(\text{length}(x))\}$$

is immune, so NL Theorem is true for every threshold function f in T.

The argument in this section is applicable in other fields dealing with compressible data. A concrete example is the problem of distinguishing between an original picture and a slightly "photoshopped" version of it.

Chapter 5

The Halting Problem and Beyond

5.1 Computability theory revisited

By the end of the nineteenth century, mathematicians started to be interested more and more in formalising the main mathematical constructs, definitions, statements and proofs. In 1889, G. Peano defined axiomatically the natural numbers $1, 2, \ldots, n$ using the induction axiom.[1] Peano axioms became the standard definition of natural numbers, and the symbols he introduced for the basic set-theoretical operations $\in, \subset, \cup, \cap, \setminus$ remained a pervasive mathematical notation. See more in Section 6.1.

Skolem [401] introduced the primitive recursive arithmetic using objects defined by primitive recursion, a scheme used later by Gödel [194], Peter [317] and Kleene [247] to define the class of *primitive recursive functions*. In programming terminology, these are the functions computable by programs for which an upper bound on the number of iterations of every loop is known before entering the loop. Examples of such functions include the arithmetical operations of addition, exponentiation, multiplication, maximum/minimum.

In a series of lectures presented in Princeton in 1934, Gödel [148, pp. 39–74] (crediting in part an unpublished suggestion by J. Herbrand [248, p. 274]), introduced the larger class of *general recursive functions*, currently called *partially computable functions*: this class is the smallest class of functions from naturals to naturals (possibly with more than one argument), including all constant functions, projections, the successor function, and which is closed under functional composition, primitive recursion, and minimisation.[2] This class was proven to be equal to the following

[1]In his treatise *Arithmetices principia, nova methodo exposita*, based on R. Dedekind work on infinite sets and irrational numbers.

[2]Minimisation here is unbounded: if only bounded minimisation is used that one gets

classes of functions: Church λ-definable functions [126][3], Post computable functions [324] and Turing computable functions [398]. It is important to note that the class of partially computable functions is defined in an "unconstructive" manner, leading to strange examples of computable functions that "cannot be computed". An example is the function $f_{\text{RH}} : \mathbf{N} \to \{0,1\}$ defined by

$$f_{\text{RH}}(n) = \begin{cases} 0, & \text{if Riemann Hypothesis is true,} \\ 1, & \text{otherwise,} \end{cases}$$

which is computable (because f_{RH} is a constant function), but nobody has (yet) calculated $f_{\text{RH}}(0)$. It is conceivable that one day Riemann Hypothesis[4] will be one day proven, so $f_{\text{RH}}(0)$ will be computed. Can we replace the Riemann Hypothesis in the definition of f_{RH} with another mathematical statement such that the corresponding function will *never* be computed?

5.2 Turing machines

A Turing machine is a mathematical model of effective computation or computation by algorithms. According to [399], a Turing machine is a mathematical abstraction which has

> an unlimited memory capacity obtained in the form of an infinite tape marked out into squares, on each of which a symbol could be printed. At any moment there is one symbol in the machine; it is called the scanned symbol. The machine can alter the scanned symbol, and its behavior is in part determined by that symbol, but the symbols on the tape elsewhere do not affect the behavior of the machine. However, the tape can be moved back and forth through the machine, this being one of the elementary operations of the machine. Any symbol on the tape may therefore eventually have an innings.

A Turing machine operates with a finite alphabet of symbols, a "head" and a "finite table" of user-specified instructions (the program) and can i) write a symbol in the scanned square, ii) move the head left or right, and iii) then proceed to a subsequent instruction or halt the computation. Of

the class of primitive recursive functions.

[3]Proven independently by Church and his student S. C. Kleene.

[4]Discussed in Section 3.6.

course, Turing machines can halt or not, depending on the program and input.

Fig. 5.1: Turing Machine, reconstructed by Mike Davey as seen at Go Ask ALICE at Harvard University[5]

A (real) number is (Turing) *computable* if there exists a Turing machine which computes an arbitrarily precise approximation to that number.[6] More precisely, a real is computable if there is a Turing machine which, given an integer $n \geq 1$ on its initial tape, terminates its computation and writes the n-th digit (in some base) on its tape [302, p. 159].

The key ingredients of the definition are (1) that some n is specified at the start, (2) for every n, the computation stops in a finite number of steps, after which the machine writes the desired output.

All algebraic numbers (roots of polynomials with algebraic coefficients) and the usual transcendental reals, such as π or e, are computable. Still, not all reals are computable, as a simple argument based on cardinality shows (the set of Turing machines is countable, but the set of reals is uncountable). The number Ω is also incomputable. All real numbers that appear in natural sciences are computable: the reason for this remains an interesting *open problem* in science and philosophy of science.

[5]Source: By Rocky Acosta – Own work, `commons.wikimedia.org/w/index.php?curid=24369879`.

[6]This definition was proposed in 1912 by É. Borel [311, p. 8] using an intuitive notion of computability.

Despite their simplicity, Turing machines are capable of running very complicated computations. A universal Turing machine (UTM) – first defined and constructed by Turing [397] – is a Turing machine that can simulate an arbitrary Turing machine on every arbitrary input; see more in Figure 5.2.

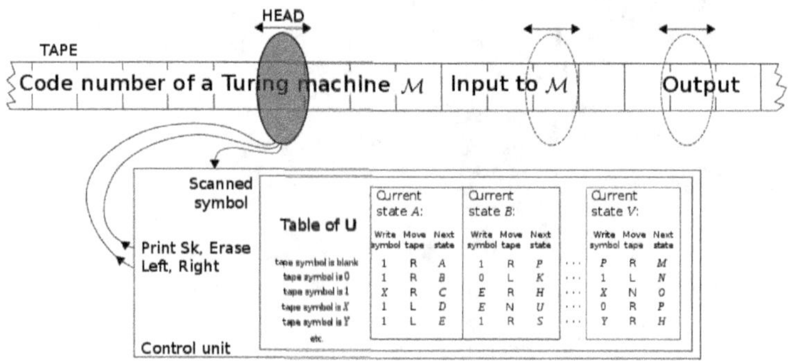

Fig. 5.2: A universal Turing machine[7]

This is achieved by reading both the code/description of the machine to be simulated and the input. In formulæ, if T is a Turing machine, $\langle T \rangle$ denotes its description, and T can have as input its own description, i.e. $T(\langle T \rangle)$.

The UTM is considered to be the origin of a stored-program computer used by J. von Neumann in the report [57]; see also [151]. Assuming the Church-Turing's Thesis, see Section 1.12, the following consequences follow: i) "anything" which can be "computed" can also be computed by a UTM and ii) conversely, any problem that is not computable by a UTM is incomputable.

5.3 More about loops and non-halting

In Section 1.6, we showed the PROGRAM(3) never ends and does not have an infinite loop. Now, we give a proof showing the existence of such a program by relating it to the Halting Problem.

[7]Source: In Wikipedia. `en.wikipedia.org/wiki/File:Universal_Turing_machine.svg`.

Assume for a contradiction that every program that never halts loops eventually. This means that for every program M and input x, one of the following two alternatives will eventually happen:

(a) $M(x)$ halts,
(b) $M(x)$ eventually repeats a sequence of instructions.

In (b), the word "eventually" means that after a finite but arbitrarily large number of instructions, a sequence of instructions repeats. Does this fact cause a problem? The answer is negative: indeed, if true, the Halting Problem could be decided by the following algorithm. Given a Turing machine M and an input x, simulate $M(x)$ and write out at every step, i) the contents of the tape, ii) the current state, and iii) the current tape position. If this configuration is duplicated, output "does not halt"; otherwise, output "halts". Since (a) or (b) are guaranteed to happen eventually, this process always terminates.[8] We have contradicted The Halting Theorem, an impossibility.

5.4 Decidability Theorem: a formal proof

In this section we prove the Decidability Theorem in Section 1.15: a set S is decidable if and only if S and its complement $\overline{S}$ are computably enumerable.

Every decidable set is computably enumerable: indeed, an enumerator can generate all positive integers $i = 1, 2, \ldots$ and print in the list those i's that the decider accepted.

If S is decidable, then its complement $\overline{S}$ is also decidable because the decider for $\overline{S}$ flips the decision of the decider for S.

Finally, if S and $\overline{S}$ are both computably enumerable by the enumerators $Enum(S)$ and $Enum(\overline{S})$, then the decider for S proceeds on an input s as follows:

(1) start the enumerators $Enum(S)$ and $Enum(\overline{S})$ to generate their lists,
(2) if s appears in the list of $Enum(S)$, then return "yes" and stops.
(3) if s appears in the list of $Enum(\overline{S})$, then return "no" and stops.

The decider stops in a finite time on every s because the union of the list produced by $Enum(S)$ and $Enum(\overline{S})$ includes all positive integers, so

[8]Note that we cannot give a bound, in terms of M and x, for the time the algorithm will run till stopping.

s will eventually appear in one of these lists. The decision of the decider is always correct because $Enum(S)$ and $Enum(\overline{S})$ have no common elements.

5.5 A finite version of the Halting Theorem

We prove the following finite version of the Halting Theorem, showing that even finite sets can be undecidable!

Theorem. *There exists $N > 0$ such that for every $n \geq N$ the finite set*

$$S_n = \{(P, v) \mid P(v)\,\text{halts}, \text{length}(P) \leq n, \text{length}(v) \leq n\},$$

is undecidable.

First, we prove that there exists $N > 0$ such that S_N is undecidable. Indeed, if for every $n > 0$, S_n is decidable, then we contradict the Halting Theorem: given (P, v), we note that $(P, v) \in S_{\max\{\text{length}(P),\text{length}(v)\}}$, which is decidable: a contradiction.

Next, assume that S_N is undecidable and $n \geq N$. Take an arbitrary $(P, v) \in S_N \subset S_n$. If S_n would be decidable, then statement "$P(v)$ halts" would be decidable in S_n, contradicting the first part of the proof.

5.6 The Acceptance Theorem

An *acceptor* is a Turing machine which, on any input w, runs a computation which a) stops in finite time and "accepts" or "rejects" w, or b) never stops. Such a Turing machine T accepts the input w if $T(w)$ halts and accepts w.

The Acceptance Problem, a closely related problem to the Halting Problem,[9] can be formulated as follows: decide for an arbitrary input pair consisting of a Turing machine T and input w whether T accepts w.

Acceptance Theorem. *The Acceptance Problem is undecidable.*

We use the notation $\langle T \rangle$ to denote the Turing machine T description; see more in Section 5.1.

The following proof uses a classical Cantor's diagonalisation argument. Let us assume by contradiction that the Acceptance Problem is decidable and H is a decider for it. On input $\langle T, w \rangle$, where T is a Turing machine and w is a string, H halts and accepts if and only if T accepts w:

[9] Challenge for the reader: What is the difference?

$$H(\langle T\rangle, w) = \begin{cases} accept, \text{ if } T \text{ accepts } w, \\ reject, \text{ otherwise}. \end{cases}$$

Note that $H(\langle T\rangle, w) = reject$ if and only if T does not accept w, i.e. $T(w)$ does not halt or $T(w)$ halts and $T(w) = reject$.

Next, we construct a new Turing machine D using the Turing machine H:

```
PROGRAM(D)
read ⟨T⟩
if H(⟨T⟩, ⟨T⟩) = accept then reject; otherwise accept
```

For every T we have:

$$D(\langle T\rangle) = \begin{cases} accept, \text{ if } T \text{ does not accept } \langle T\rangle, \\ reject, \text{ if } T \text{ accepts } \langle T\rangle. \end{cases}$$

Finally, for $T = D$ we get:

$$D(\langle D\rangle) = \begin{cases} accept, \text{ if } D \text{ does not accept } \langle D\rangle, \\ reject, \text{ if } D \text{ accepts } \langle D\rangle. \end{cases}$$

We reached a contradiction, hence neither D nor H can exist.

Note that running a Turing machine on its own description, $D(\langle D\rangle)$, is legitimate because $\langle D\rangle$ is a string.

5.7 Rice Theorem – a formal proof

Syntactic properties of programs (as, for example, the program contains an if-then-else statement) are decidable.

What about semantic properties, like the program halts for all inputs? In this section, we prove a more general result than the Halting Theorem, which was used in Section 4.5, stating that all non-trivial semantic properties of programs are undecidable.

A property P of programs is identified with the set of all programs satisfying P and vice-versa, a set of programs defines a property of programs.

> **Rice Theorem.** [339] *Let P be a property of programs fulfiling the following two conditions:*
>
> *(1) P is non-trivial; that is, some but not all programs have P,*

(2) P is a semantic property; that is, a property of the sets of program outputs.

Then, P is undecidable.

For a proof, we assume for the sake of a contradiction that P is decidable and let R_P be a decider for P. We show that R_P can be used to decide the Acceptance Problem, which contradicts the Acceptance Theorem in Section 5.6.

First, let $T_\emptyset$ be a program which accepts no input and assume without loss of generality that $\langle T_\emptyset \rangle \notin P$ (otherwise proceed with the complement of P instead of P).

Second, because P is non-trivial, there exists a program T with $\langle T \rangle \in P$.

Third, the following program constructs a new program $N_{M,w,T}$ – based on arbitrary M, w, and T – as follows:

```
PROGRAM(Construct the program N)
read M
read w
read T
read x
construct N_{M,w,T}
run M(w)
    if M(w) does not accept then N_{M,w,T}(x) = M(w)
    otherwise N_{M,w,T}(x) = T(x)
stop
```

Note that for every x, $N_{M,w,T}(x) = T(x)$ if and only if $M(w)$ accepts.

Next, we construct a decider for the Acceptance Problem, contradicting the Acceptance Theorem.

```
PROGRAM(Decide acceptance)
read M
read w
    construct N_{M,w,T}
        if R_P(⟨N_{M,w,T}⟩) = 1 accept and stop
        otherwise reject
```

We have the following formula for every x:

$$N_{M,w,T}(x) = \begin{cases} T(x), & \text{if } M(w) \textit{ accepts}, \\ M(w), & \text{otherwise}. \end{cases}$$

If $M(w) = accept$, then for all $x, N_{M,w,T}(x) = T(x) \in P$, hence we have $R_P(\langle N_{M,w,T}\rangle) = 1$. Conversely, if $M(w)$ does not accept, then for all $x, N_{M,w,T}(x) = M(w) \notin P$, hence $R_P(\langle N_{M,w,T}\rangle) = 0$.

Here are more examples[10] of semantic properties given by their corresponding sets:

(1) The set of programs returning 0 for every input.
(2) The set of programs returning 0 for at least one input.
(3) The set of constant programs.
(4) The set of programs that are undefined for some input.
(5) The set of programs that stop on all inputs.

All complements of the above sets are also incomputable.

5.8 Specker's Theorem

The definition (2.4) of Ω invites a short discussion of the computability of real numbers.

We start with a "nice" computable enumeration of all rationals [420]:

$$r_0 = 0, r_{2n} = 1 + r_n, \, n > 0, r_{2n+1} = \frac{1}{(1+r_n)}, \, n \geq 0.$$

So,

$$r_0 = 0, r_1 = \frac{1}{(1+r_0)} = 1, r_2 = 1 + r_1 = 2, r_3 = \frac{1}{(1+r_2)} = \frac{1}{2}, \ldots$$

The adjective "computable" means that the function $n \longmapsto r_n$ can be computed by two computable functions f^1, f^2 such that $r_n = \frac{f^1(n)}{f^2(n)}$. The sequence $(r_{2n+1})_n$ enumerates without repetitions all rational numbers in the interval $(0, 1]$. One can easily extend this enumeration to a computable enumeration R_n of all rationals. A rational number is "given" by its unique index n in the enumeration R_n.[11] In this way, every sequence of rationals corresponds to a unique sequence of naturals, and vice-versa; hence the computability and complexity theories can be extended naturally to the set of rationals. In particular, we can talk about computable and computably enumerable sets of rationals. In the same way, we can study the computability of subsets of many countable sets, for example, the set of graphs.

[10] See Section 4.5 for other examples.
[11] Challenge for the reader: Why?

A computably enumerable set of rationals $a_1, a_2, \ldots a_n, \ldots$ is *computably convergent* if there is a computable convergence function G (from naturals to naturals) such that for every natural $n, m, k > 0$, the inequality

$$|a_n - a_m| < \frac{1}{k},$$

whenever $n, m \geq G(k)$. A real number is *computable* if it is the limit of a computably convergent sequence of rationals. The well-known formulæ

$$e = 1 + \frac{1}{1!} + \frac{1}{2!} + \frac{1}{3!} + \ldots,$$

$$\pi = 4\left(1 - \frac{1}{3} + \frac{1}{5} - \frac{1}{7} + \ldots\right),$$

show that e and π are computable reals.

Fig. 5.3: E. Specker[12]

Is every real number computable? The reader may remember that this question was negatively answered in Section 5.1 relative to a different, but equivalent, definition of a computable real number.

More insight into the "structure" of incomputable reals is given by:

[12]Source: In Wikipedia. `en.wikipedia.org/wiki/File:Ernst_Specker.jpeg`.

Specker Theorem. [375] *There exists a computably enumerable sequence of rationals*

$$0 < s_1 < s_2 \cdots < s_n \cdots < 1, \tag{5.1}$$

which does not computably converge, hence $s = lim_{n\to\infty} s_n$ *is not a computable real.*

To prove this result, we pick a computably enumerable set of natural numbers A which is not decidable; for example, one can take the domain of a universal Turing machine.[13] It follows that A is infinite,[14] hence we can choose an one-to-one computable enumeration $(a_i)_{i\geq 0}$ of A. Next, following [60, p. 196], we define the sequence (s_n) of rationals as follows:

$$s_n = \sum_{i=0}^{n} 2^{-a_i-1}. \tag{5.2}$$

Each s_n is positive and rational, and the sequence s_n is strictly increasing. As (a_i) is a computable enumeration of A without repetitions, we have

$$0 < s_n < \sum_{i=0}^{\infty} 2^{-i-1} = 1.$$

According to a classical calculus result, the sequence (s_n) has a supremum s. To finish, we assume by contradiction that the sequence s_n is computably convergent and the computable function g is the convergence function. We construct the computable function $G(n) = g(2^{n+1})$ and note that for all naturals n, m with $n > G(m)$, we have:

$$s_n - s_{G(m)} = |s_n - s_{G(m)}| < 2^{-m-1}. \tag{5.3}$$

From (5.2) and (5.3), for all n, m we have:

$$n > G(m) \Rightarrow a_n > m. \tag{5.4}$$

According to (5.4), we obtain the equivalence

$$m \in A \iff a_n = m, \text{ for some } n \in \{0, 1, 2, \ldots, G(m)\},$$

which contradicts the hypothesis that A is not computable.[15]

[13] Challenge for the reader: Prove this statement.

[14] Challenge for the reader: Why?

[15] Challenge for the reader: Is there any similarity between the number Ω and the real number s in the Specker Theorem?

Specker Theorem is fundamental both in constructive mathematics and traditional mathematical analysis. Specker sequences are counter-examples that contradict, in the context of the most natural algorithmic conception of a constructive continuum, fundamental principles such as the Weierstrass Theorem on the existence of a limit for a bounded monotone sequence and the Bolzano-Weierstrass Theorem on the choice of a converging subsequence. In traditional mathematics, Specker Theorem reveals simple objects with a deep complex nature.

5.9 Can incomputability be "experimentally tested"?

Why should we be interested in answering the above question? After all, incomputability is established by mathematical proof, so why would we need experimental corroboration, a weaker argument? An example is a random number generator certified (by a mathematical proof) always to produce an incomputable infinite sequence of random bits. Indeed, the mathematical proof certifying incomputability is part of a mathematical model which uses certain physical assumptions: its veracity rests on those assumptions. The fact that each individual assumption is reasonable does not automatically guarantee that globally the set of assumptions is also reasonable. Experimental testing is essential not only for corroborating the conclusion of the proof but also for supporting the adequacy of the model. Furthermore, thorough testing allows one to detect any issues with assumptions made in the theoretical analysis of a device or its practical deployment.

Can we test incomputability with a statistical test, that is, with a method of statistical inference to decide whether the data at hand sufficiently support a particular hypothesis? The answer is *negative*. Intuitively, this is a consequence of the "asymptotic" nature of computability and its negation: *finite variations do not change them*. For example, if the sequence $x_1x_2\dots x_n\dots$ is computable (incomputable), then the sequences $y_1y_2\dots y_mx_1x_2\dots x_n\dots$ and $x_kx_{k+1}\dots x_m\dots$ are also computable (incomputable) for every string $y_1y_2\dots y_m$ and positive integer k.[16] For example, the Champernowne binary sequence [123]

$$0, 1, 00, 01, 10, 11, 000, \dots \tag{5.5}$$

obtained by concatenating all non-empty binary strings in shortlex order is computable[17] and *normal*, i.e. its digits are uniformly distributed: all digits

[16] Challenge for the reader: Why?

[17] Challenge for the reader: Why?

are equally likely, all pairs of digits are equally likely, all triplets of digits are equally likely, and so on. Normality is a "symptom" of randomness, and computability is a "symptom" of non-randomness.[18] The Champernowne sequence shows that these symptoms can be compatible; no statistical test can detect its computability, hence non-randomness.

Does this mean that incomputability cannot be "experimentally tested"? Of course, no. The first tests have been proposed in [53, 88]. In what follows, we will describe such a test used in assessing the quality of outputs of quantum random generators [88, 3, 245].

We continue with a topic apparently unrelated to the question discussed in this section: testing primality of positive integers. Primality is considered computationally *easy* because polynomial-time algorithms (in the size of the input) solve it; the first such algorithm was proposed in 2004 [13].[19] However, every known primality polynomial-time algorithm is "practically slow",[20] so probabilistic algorithms for primality are instead used [301, 332, 371, 372].[21]

To test the primality of a positive integer n, the Solovay-Strassen primality test generates the first k natural numbers uniformly distributed between 1 and $n-1$, inclusive, and, for each $i \in \{i_1, \ldots, i_k\}$ and checks "quickly" the validity of a predicate $W(i, n)$ (called the Solovay-Strassen predicate). If $W(i, n)$ is true, then "i is a witness of n's compositeness"; hence n is certainly not prime. Otherwise, the test is inconclusive,[22] but in this case, the probability that n is prime is greater than $1 - 2^{-k}$. This result is based on the fact that *at least half* the i's between 1 and $n-1$ satisfy $W(i, n)$ if n is composite, and *none* of them satisfy $W(i, n)$ if n is prime [373].

Chaitin and Schwartz [121] showed that if c is a large enough positive integer and s is a long enough c-random binary string[23], then n is prime if and only if $Z(s, n)$ is true, where Z is a predicate constructed directly from $O(\log n)$ conjunctions of negations of W predicate.

---

[18] Turing and Champernowne have been friends since they were undergraduates at King's College, Cambridge. Champernowne's work could have stimulated Turing's interest in the constructive definition of real numbers, the "seed" for his famous article [397].

[19] In contrast, factorisation of positive integers is "thought", but not proven, to be a computationally *difficult* problem. Currently, one cannot factorise a positive integer of 500 decimal digits that is the product of two randomly chosen prime numbers. This fact is exploited in the RSA cryptosystem implementing public-key cryptography [340].

[20] Challenge for the reader: Justify this claim.

[21] Currently the best runs in time $O((\log\ n)^6)$.

[22] Honestly, the answer should be "don't know".

[23] Recall that a string s is c-random if $K(s) \geq \text{length}(s) - c$.

In detail, we first define the Solovay-Strassen predicate $W(i,n)$ by

$$\left(\frac{i}{n}\right) \neq i^{(n-1)/2} \mod n,$$

where $\left(\frac{i}{n}\right)$ is the Jacobi symbol[24] with $i \in \mathbf{N}, i < n-1$.

Consider $s = s_0 \dots s_{m-1}$ a binary string (of length m) and n an integer greater than 2. Let k be the smallest integer such that $(n-1)^{k+1} > 2^m - 1$; we can thus rewrite the number whose binary representation is s into base $n-1$ and obtain the unique string $d_k d_{k-1} \dots d_0$ over the alphabet $\{0, 1, \dots, n-2\}$, that is,

$$\sum_{i=0}^{k} d_i (n-1)^i = \sum_{t=0}^{m-1} s_t 2^t.$$

The predicate $Z(s,n)$ is defined by

$$Z(s,n) = \neg W(1+d_0, n) \wedge \cdots \wedge \neg W(1+d_{k-1}, n), \tag{5.6}$$

where W is the Solovay-Strassen predicate.

The digits of s (rewritten in base $n-1$) define the witnesses used to test the primality of n.

> **Chaitin-Schwartz Theorem.** [121] *For all sufficiently large c, if s is a c-random string of length $(l+2c)$ and n is an integer whose binary representation is l bits long, then $Z(s,n)$ is true if and only if n is prime.*

This result cannot be used to de-randomise[25] Solovay-Strassen probabilistic algorithm because the set of c-random strings is incomputable.[26] However, the result can be used to test the quality of long binary strings by comparing their "computational power" with that of c-random strings of the same length.

Following [4], we test the primality of Carmichael numbers, which are composite positive integers n satisfying the congruence $b^{n-1} \equiv 1 \pmod{n}$ for all integers b relatively prime to n. A Carmichael number passes a Fermat primality test [141, Section 31.8] to every base relatively prime to the

[24] If the prime factorisation of the odd number n is $p_1^{a_1} p_2^{a_2} \dots p_k^{a_k}$, then $\left(\frac{i}{n}\right) = \left(\frac{i}{p_1}\right)^{a_1} \left(\frac{i}{p_2}\right)^{a_2} \dots \left(\frac{i}{p_k}\right)^{a_k}$.

[25] That is, to transform the probabilistic algorithm into an equivalent deterministic algorithm.

[26] In fact, highly incomputable [61]: no infinite set of c-random strings is computable.

number, but few pass the Solovay-Strassen test. Increasingly, Carmichael numbers become "rare".[27]

Standard statistical randomness tests focus on properties of the distribution of bits or bit strings within sequences, failing to distinguish between pseudo-random number generators and quantum random number generators. To address this issue, in [4], the ability of random strings to de-randomise the Solovay-Strassen probabilistic test of primality was used to compare the algorithmic randomness of strings generated by a QRNG and different PRNGs. Despite leading to mostly inconclusive results, the tests showed some advantages offered by a 3D-QRNG against PRNGs with respect to the randomness of its outputs.

The following test, the fourth Chaitin-Schwartz-Solovay-Strassen test (CSS4) in [4], showed the highest potential for distinguishing between sources of random strings. Recall that the crucial fact is that the set of c-random strings is highly incomputable.

We construct the Chaitin-Schwartz predicate $Z(s, n)$ from (5.6) and generate a pool of "witnesses" composed of the digits s in base $n-1$. Then, we fix $c = l - 1$ where l is the l-bit binary representation of n and sample s from chunks of $l(l+2c)$ bits. The criterion is the number of direct violations of the Chaitin-Schwartz Theorem generated by a set of extracted witnesses.

In [3], Carmichael numbers were used in the majority of the tests. However, despite Carmichael numbers having a larger $L_{ss}(n)$, it is difficult to find *Z-liars* due to the length of their binary representation. For example, for the smallest Carmichael number, more than 70×2^{32} bits would need to be read to find a *Z-liar* since the Solovay-Strassen test guarantees a predicate is true with a probability of at least one-half when n is a composite number. For smaller numbers, we expect to see a larger number of *Z-liars*. Thus, for this test, only odd composite numbers less than 50 were used for each round, and the process was repeatedly parsed through each string with an incremental bit offset.

In [245], a similar approach was taken by applying these tests to a different set of PRNGs and two different QRNGs with a larger set of numbers; each string tested had a length of 2^{26}. Once again, the QRNGs showed no clear advantage over the PRNGs. Moreover, the difficulty of finding *Z-liars* led to a similar limitation in terms of numbers tested; *Z-liars* were only observed for composites $n \leq 25$. Still, an essential characteristic of this

[27] There are 1,401,644 Carmichael numbers in the interval $[1, 10^{18}]$.

test was confirmed: its sensitivity to the size of the pool of unique bases extracted from the random strings. No *Z-liars* were recorded when a repetitive structure generated by their sampling process was present. For this reason, a variation of this test was performed.

We tested two PRNGs and two QRNGs: the hashing function SHA3, considered a "cryptographically secure PRNG", the Python3 Mersenne Twister-based generator, and a 3D-QRNG described in Section 4.12.

Since the number of Solovay-Strassen tests increases with longer binary representations, the probability of observing a *Z-liar* becomes smaller, so a large pool of unique bases was required to detect a significant number of *Z-liars* [3]. Thus, ten sets of strings of size 2^{32} were produced with each generator and the shifting process described in [3] was applied for the test.

	9	15	21	25	27	33	35	39	45	49
SHA3	265.6	60.3	0	0.2	0	0	0	0	0	0
Python3	260.1	58	0	0.3	0	0	0	0	0	0
Qutrits (2021)	536.4	131.9	0	0.2	0	0	0	0	0	0

Table 5.1: Average number of violations by composite (over ten strings)

The average number of *Z-liars* over the composite numbers less than 50 was taken as the metric. Despite only detecting *Z-liars* for composites up to 25, there was a noticeable difference between sources for the numbers 9 and 15. For these numbers, from our predicate construction, a minimum of 40×2^{13} bits and 40×2^{10} bits were needed for a c-random string to have a chance of finding a *Z-liar*.

As noted in Section 4.10, patterns in long enough sequences of random events are inevitable. Since a lower quality of randomness increases the rate at which this occurs, the gap between the number of unique witnesses extractable between RNGs with different qualities of randomness widens. Thus, given long enough strings, we have better chances to observe this behaviour. Since many unique bases are required to increase the likelihood of finding *Z-liars*, from Table 5.1 and Figure 5.4 we see the advantage offered by a 3D-QRNG generator over other alternative sources of randomness.

Is this analysis enough? The answer is no! We have to show that the difference between the 3D-QRNG and the PRNGs is *statistically significant*.

Almost all of the tests indicating statistically significant results assume a known probability distribution, like a normal distribution. Unfortunately, in this case, no probability distribution is known, so such an assumption would be wrong. This is the reason the non-parametric and distribution-free two-sample Kolmogorov-Smirnov test [133] was used.

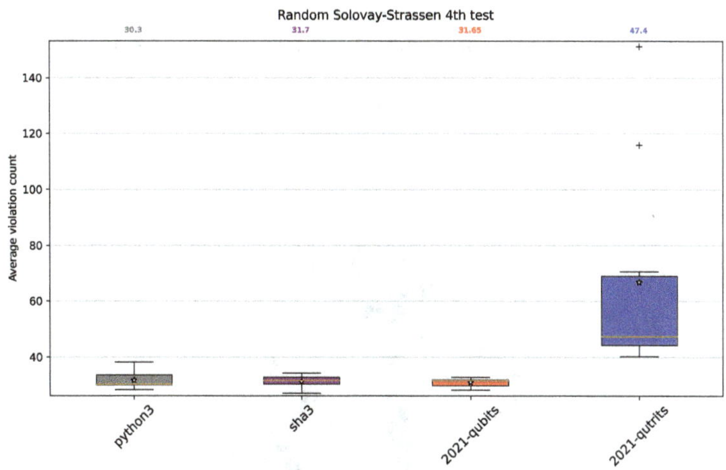

Fig. 5.4: Fourth Chaitin-Schwartz-Solovay-Strassen test: distribution of the average counts of violations of the Chaitin-Schwartz Theorem for all odd composite numbers less than 50

The Kolmogorov-Smirnov test[28] identifies if the two datasets differ significantly without any prior assumption about an underlying distribution. The test answers the question, "How likely is it that we would see two sets of samples if they were drawn from the same (but unknown) probability distribution?".

According to the Kolmogorov-Smirnov test, two datasets are statistically significant if the p-value obtained is less than a given small value, chosen to reduce the chance of false positives.

The Kolmogorov-Smirnov test with the commonly used critical p-value of 0.05 indicated a smaller but significant difference between the 3D-QRNG and PRNGs in [4].

With the much smaller p-value of 0.005, the analysis in [16] showed a clear advantage of 3D-QRNGs over PRNGs; see Table 5.2.

[28]Named after A. Kolmogorov and N. Smirnov.

	SHA3	qutrits
Python3	0.9780	0.0047
SHA3		0.0047

Table 5.2: Kolmogorov-Smirnov test p-values for the fourth Chaitin-Schwartz-Solovay-Strassen test with the Z-liar count metric

Fig. 5.5: A. N. Kolmogorov[29]

This example shows that incomputability, an essential property for evaluating the quality of random number generators, can be algorithmically tested.

5.10 How "undecidable" is the Halting Problem?

The Halting Problem is undecidable, but can it be solved for "most", or, even better, "almost all", inputs?

This question is natural but challenging to formulate rigorously because we have to formalise the adjective "most". There are various ways to do it, quantitatively or quantitatively.

[29]Source: In Wikipedia. en.wikipedia.org/wiki/File:Andrej_Nikolajewitsch_Kolmogorov.jpg.

We have shown in Section 1.16 approximate solutions to Halting Problem which "solve" *any* instance of the problem with a pre-given probability. Hamkins and Miasnikov [211] have shown that the Halting Problem for Turing machines with one-way tape is "easy" decidable on a set of asymptotic probability one – "almost everywhere" in probabilistic terms. Here are some details. The asymptotic density (or probability) of a set X of Turing machines is defined as follows:

$$Prob(X) = \lim_{n\to\infty} \frac{\#(X \cap P_n)}{\#(P_n)},$$

where P_n is the set of all n-state Turing machines. What does it mean that a set X has probability one, $Prob(X) = 1$?[30] Given a natural number $0 < p < 100$ there exists sufficiently large n (which depends on p) such that more than $p\%$ of all n-state programs are in X. The value of p can be chosen as close to 100% as desired (but, of course, not 100).

The main result is:

> **Hamkins-Miasnikov Theorem.** *The Halting Problem for Turing machines with one-way tape is polynomial-time decidable on a set with asymptotic probability one.*

The proof in [211] uses a random walk with an equal likelihood of moving left or right on a one-way infinite tape.

Furthermore, A Rybalov [351] proved that the Halting Problem is undecidable when restricted to an arbitrary "very large" set of Turing machines.[31] Other results regarding the "size" of undecidability of the Halting Problem have been obtained in [86, 45, 44].

5.11 Most programs stop quickly or never halt

The common wisdom says that writing short programs that stop after a very long time is possible. However, it is less evident that there are only a few such programs; these programs are "exceptions".

To understand the phenomenon, we start by considering all programs of length $N = 3$ for a simple Turing machine M whose domain is the finite set $\{000, 010, 011, 100, 110, 111\}$. The halting "history" of these programs, presented in Table 5.3, shows the times the programs in the domain of M

[30]Sometimes, this is phrased as "almost all Turing machines are in X".

[31]Technically, strongly generic set.

halt. The program $M(p_1)$ halts at time $t = 1$, so it is indicated by an "h" on the row corresponding to p_1 and time $t = 1$; the program $M(p_4)$ hasn't halted at time $t = 5$, so, on the row corresponding to p_4 and time $t = 4$, there is a blank. Finally, programs which haven't stopped by time $t = 17$, never stop.

Program	$t = 1$	$t = 2$	$t = 5$	$t = 8$	...	$t = 14$	$t = 15$	$t = 16$	$t = 17$
$p_1 = 000$	h	h	h	h	h	h	h	h	h
$p_2 = 001$									
$p_3 = 010$							h	h	h
$p_4 = 011$				h	h	h	h	h	h
$p_5 = 100$						h	h	h	h
$p_6 = 101$									
$p_7 = 110$	h	h	h	h	h	h	h	h	h
$p_8 = 111$								h	h

Table 5.3: Halting "history" for 3-bit programs

It is worth noting the following facts in Table 5.3:

- the program $M(p_1)$ halts exactly at time $t = 1$,
- the set of 3-bit programs which halt exactly at time $t = 1$ consists of $\{p_1, p_7\}$, so the 'probability' that a randomly chosen 3-bit program halts at time $t = 1$ is

$$\#\{\text{3-bit programs halting at time 1}\}/\#\{\text{3-bit programs}\}$$

$$= \#\{p1, p7\}/8 = 2/8 = 1/4.$$

- the set of 3-bit programs which halt by time $t = 8$ consists of $\{p_1, p_4, p_7\}$, so the 'probability' that a randomly picked 3-bit program halts by time $t = 8$ is #{3-bit programs halting by time 8}/#{3-bit programs} $= \#\{p1, p4, p7\}/8 = 3/8$,
- the 'probability' that a random 3-bit program eventually stops is #{3-bit programs that halt}/#{3-bit programs} $= 6/8$,
- the program $M(p_4)$ hasn't stopped by time $t = 5$, but stops at time $t = 8$,
- the 'probability' that a 3-bit program does not stop by time $t = 5$ and that it eventually halts is #{3-bit programs that eventually halt that have not stopped by time $t = 5$}/#{3-bit programs} $= \#\{p3, p4, p5, p8\}/8 = 4/8 = 1/2$,

- the 'probability' that a 3-bit program eventually stops given that it has not halted by time $t = 5$ is $\#\{$3-bit programs that eventually halt that have not stopped by time $t = 5\}/\#\{$3-bit programs that have not halted by time $t = 5\} = 4/(8-2) = 2/3$,
- the 'probability' that a 3-bit program halts at time $t = 8$ given that it has not halted by time $t = 5$ is $\#\{$3-bit programs that halt by time $t = 8$ but not by time $t = 5\}/\#\{$3-bit programs that have not halted by time $t = 5\} = 1/6$.

There are two approaches for calculating the probability that an N-bit program will halt. The first approach, initiated by Chaitin [114], is to have the probability space range only over programs; this is the approach taken when computing the Omega number defined in (2.4). In that case, the probability is

$$Prob_N = \#\{p \in \{0,1\}^N \mid p \text{ halts}\}/2^N.$$

For a self-delimiting Turing machine,[32] $Prob_N$ goes to zero when N tends to infinity since it becomes more and more likely that any given N-bit string is an extension of a shorter halting program. For a universal non-self-delimiting Turing machine, the probability is always nonzero for large enough N. After some point, this machine will simulate a total Turing machine (one that halts on all inputs), so some fixed proportion of the space will always contribute. In this case, the probability is incomputable; in general, 1 is the best computable upper bound one can find. In this model, it matters only whether a program halts; the time at which a halting program stops is irrelevant.

Here we take a different path: the probability space extends over space and time, and we evaluate the probability that a random N-bit program that hasn't stopped by some given time will halt later. The stopping time of a halting program is paramount. As there is no uniform distribution on the integers, we must also choose a distribution on times. Not any distribution works: we select one for which most long times are rare. This gives a certain "invariance": in the limit, the distribution we choose doesn't matter much.

Here is the formalism. We fix a universal Turing machine U and a pair (N, T), where N represents the length of the program and T is the "time-window", i.e. the interval $\{1, 2, \ldots, T\}$, where the computational time is

[32] A Turing machine is self-delimiting (or prefix-free) if its domain is prefix-free, i.e. no extension of a string in the domain is itself in the domain. For example, the set $\{1^n 0 \mid n > 0\}$ is prefix-free. A detailed construction is given in Section 3.12.

observed. The probability space is thus

$$Space_{N,T} = \{0,1\}^N \times \{1,2,\ldots,T\}.$$

In what follows, both programs and times are assumed to be *uniformly distributed*.[33] For $A \subseteq Space_{N,T}$, we define $\text{Prob}_{N,T}(A)$ to be $\#(A) \cdot 2^{-N} \cdot T^{-1}$.

Further,

$$A_{N,T} = \{(p,t) \in Space_{N,T} \mid U(p) \text{ stops exactly at time } t\},$$

and

$$B_{N,T} = \{(p,t) \in Space_{N,T} \mid U(p) \text{ stops by time } t\}.$$

From $\#(A) \leq 2^N$, we derive the following inequalities:

$$\text{Prob}_{N,T}(A_{N,T}) = \frac{\#(A_{N,T})}{2^N \cdot T} \leq \frac{2^N}{2^N \cdot T} = \frac{1}{T},$$

$$\text{Prob}_{N,T}(B_{N,T}) \leq \frac{\#(B_{N,T})}{2^N \cdot T} \leq \frac{2^N \cdot T}{2^N \cdot T} = 1.$$

Recall that bin is the computable bijection between the set of positive integers and $\{0,1\}^*$ defined in (2.1).

We will work with Turing machines M which process strings into strings and with a slightly different complexity than K called the *natural complexity* [106]: for every Turing machine and every string $x \in \{0,1\}^*$,

$$\nabla_M(x) = \min\{n \geq 1 \mid M(\text{bin}(n)) = x\}.$$

Invariance Theorem. *We can effectively construct a universal machine U such that for every machine M, there is a constant $\varepsilon > 0$ (depending on U and M) such that $\nabla_U(x) \leq \varepsilon \cdot \nabla_M(x)$, for all strings x.*

For example, if $U(0^i1x) = M_i(x)$ (where (M_i) is an effective enumeration of all Turing machines), then $\nabla_U(x) \leq (2^{i+1}+1) \cdot \nabla_{M_i}(x)$, because $0^i1\text{bin}(m) = \text{bin}(2^{i+1+\lfloor \log(m) \rfloor} + m)$, for all $m \geq 1$. Note that 0^i1x codes the Turing machine M_i, 1 marks the end of the encoding, and x is the input.

[33]This is possible because the sets $\{0,1\}^N$ and $\{1,2,\ldots,T\}$ are finite.

Fig. 5.6: G. J. Chaitin[34]

In what follows, we fix a universal Turing machine U and write ∇ instead of ∇_U.

Working with self-delimiting Turing machines, Chaitin [116] has given the following estimation of the complexity of the runtime of a program that eventually halts:

Chaitin Runtime Theorem. *There is a constant c such that if $U(\text{bin}(i))$ halts in time t, then*

$$\nabla(\text{bin}(t)) \leq 2^{\text{length}(\text{bin}(i))} \cdot c \leq i \cdot c. \tag{5.7}$$

As t is the first time $U(\text{bin}(i))$ halts,[35] the above relation puts a limit on the *natural complexity* of the time t a program $\text{bin}(i)$, that eventually halts on U, has to run before it stops; this gives a limit on the time t because only finitely many strings have complexity bounded by a constant. In view of (5.7), *the bound depends only upon the length of the program*; the program itself (i.e. $\text{bin}(i)$) does not matter. In other words, all strings of the same length have the same limit.

Because $\lim_{t\to\infty} \nabla(\text{bin}(t)) = \infty$, there are only finitely many integers t satisfying the inequality (5.7). That is, there exists a critical value $T_{critical}$ (depending upon U and $\text{length}(\text{bin}(i))$) such that if for each $t < T_{critical}$, $U(\text{bin}(i))$ does not stop in time t, then $U(\text{bin}(i))$ never halts:

[34]Source: V. Chaitin, email to C. Calude on 28 April 2023.

[35]Of course, if $U(\text{bin}(i))$ halts in time t, it stops also in time $t' > t$, but only finitely many t' satisfy the inequality (5.7).

If $U(\text{bin}(i))$ does not stop in time $T_{critical}$, then $U(\text{bin}(i))$ never halts.

So, what prevents us from running the computation $U(\text{bin}(i))$ for the time $T_{critical}$ and deciding whether it halts or not? Obviously, the incomputability of $T_{critical}$. Neither the natural complexity ∇ nor any other size complexity, like K or H, is computable (see [164]). There are large integers t with small complexity $\nabla(\text{bin}(t))$, but they cannot be effectively "localised" because one cannot effectively compute a bound $b(k)$ such that $\nabla(\text{bin}(t)) > k$ whenever $t > b(k)$.

The above analysis suggests that "a program that has not stopped after running for a long time has smaller and smaller chances to stop eventually". The bound (5.7) is not computable. Still, can we "extract information" from the inequality (5.7) to derive a computable probabilistic description of this phenomenon?

Without loss of generality, we assume that the universal (non-necessarily self-delimiting) Turing machine U has a built-in counting instruction. Based on this, there is an effective transformation which, for each program p, produces a new program $time(p)$ such that there is a constant $c > 0$ (depending upon U) for which the following three conditions are satisfied:

(1) $U(p)$ halts iff $U(time(p))$ halts,
(2) $\text{length}(time(p)) \leq \text{length}(p) + c$,
(3) if $U(p)$ halts, then it halts at the step $t_p = \text{bin}^{-1}(U(time(p)))$.

Intuitively, $time(p)$ either calculates the number of steps t_p till $U(p)$ halts and prints $\text{bin}(t_p)$, or, if $U(p)$ is not defined, never halts.[36]

Theorem. [107] *If $U(p)$ stops exactly at time t_p, then*

$$\nabla(\text{bin}(t_p)) \leq 2^{\text{length}(p)+c+1}. \tag{5.8}$$

Here is the proof. First we note that for every program p of length at most N, $\text{bin}^{-1}(p) < 2^{N+1}$. Indeed, $\text{length}(p) = \text{length}(\text{bin}(\text{bin}^{-1}(p))) \leq N$ implies

$$2^{\text{length}(p)-1} \leq \text{bin}^{-1}(p) < 2^{\text{length}(p)+1} \leq 2^{N+1}. \tag{5.9}$$

Since $U(p) = U(\text{bin}(\text{bin}^{-1}(p)))$, we have:

$$\nabla(U(p)) = \min\{i \geq 1 \mid U(\text{bin}(i)) = U(p)\} \leq \text{bin}^{-1}(p),$$

[36] The constant c can be taken to be less than or equal to 2, as the counting instruction is used only once, and we need one more instruction to print its value; however, we don't need to print the value $U(p)$.

hence

$$\nabla(\text{bin}(t_p)) = \nabla(U(time(p))) \leq \text{bin}^{-1}(time(p)) < 2^{\text{length}(p)+c+1},$$

because $\text{length}(time(p)) \leq \text{length}(p) + c$ and (5.9).

We have obtained the following result:

> **Theorem.** *Given an integer $k > 0$, we can effectively compute a time bound T such that the probability that an N-bit program p will eventually halt on U, given that it has not stopped by time T, is smaller than 2^{-k}.*

Following [105], the critical time t_p can be used to yield the following classification of Turing computations:

- $U(p)$ halts in time bounded by t_p.
- $U(p)$ halts in time not bounded by t_p.
- $U(p)$ does not halt.

The last case, when $U(p)$ does not halt, can be refined in terms of space complexity. The *computational space*, $space_U(p)$ used by U on p is defined to be the number – finite or infinite – of cells used by $U(p)$ during its computation; a cell used at least once is counted as used.[37] Obviously, if $space_U(p)$ is finite, then the computation process described above can have only a finite number of different configurations.[38]

Clearly, $space_U(p) < \infty$ whenever $U(p)$ stops, and $U(p)$ does not stop if and only if $time_U(p) = \infty$. We can therefore classify input programs p according to $time_U$ and $space_U$ in the following three sets:

(1) $\{p \mid time_U(p) < \infty\}$,
(2) $\{p \mid time_U(p) = \infty, space_U(p) < \infty\}$,
(3) $\{p \mid space_U(p) = \infty\}$.

5.12 How large is the set of spurious correlations?

> When we see lines sloping together, bars rising together, or points on a scatterplot clustering, the data practically begs us to assign a

[37]This definition differs slightly from the space complexity usually employed in computational complexity theory [33], which treats the space as infinite if the time is infinite.
[38]A configuration records the current state, tape contents and head location [362, p. 140].

reason. We want to believe one exists.[39]

Is this belief correct or just reasonable? The existence of spurious correlations shows that the answer is negative. But what is a spurious correlation?

According to the *Oxford Dictionary* (2015), the adjective spurious means

> "Not being what it purports to be; false or fake. False, although seeming to be genuine. Based on false ideas or ways of thinking."

The above definition is semantic, hence it depends on an assumed theory: one correlation can be spurious according to one theory but meaningful with respect to another. To answer the question posed in the title of this section, we need a definition independent of any theory. In what follows, we will use the definition in [99]:

> A correlation is *spurious* if it appears in a "randomly" generated database.[40]

Ramsey correlations discussed in Section 4.10 are spurious because they appear in every database; in particular in "randomly" generated databases.

We will use incompressibility discussed in Section 2.5 as a model of randomness, and here we recall the following facts:

Fix $n > 0$, so we have 2^n n-bit strings.

(1) A string x is c-compressible ($1 \leq c < \text{length}(x)$) if $K(x) \leq \text{length}(x) - c$ and at least $2^n - 2^{n-c+1} + 1$ strings of length n are incompressible by c.

(2) The probability that a string x of length n is not c-compressible is

$$\#\{x \mid \text{length}(x) = n, K(x) > n - c\} \cdot 2^{-n} \geq 1 - 2^{1-c} + 2^{-n}.$$

Take now $c = n/2$, a mild compression degree. Then, the probability that a string x of length n is not $n/2$-compressible is

$$\begin{aligned}\#\{x \mid \text{length}(x) = n, K(x) \geq n - n/2\} \cdot 2^{-n} \\ = \#\{x \mid \text{length}(x) = n, K(x) \geq n/2\} \cdot 2^{-n} \\ \geq 1 - 2^{1-n/2} + 2^{-n} \longrightarrow_{n\to\infty} 1.\end{aligned}$$

We can now give the following definition:

---

[39] `hbr.org/2015/06/beware-spurious-correlations`.

[40] We note that this correlation can appear in other databases with "meaningful" entries.

(1) A k-correlation in an n-bit-string x is a substring of length k of the string x. (There are $n - k + 1$ possible correlations in x.)

(2) A correlation is *spurious* if it appears in a c-incompressible n-bit-string, for some $c > 0$.

We note that the definition of correlation depends on the size of the string, i.e. the size of the database. Ramsey correlations are an example.

Do non-spurious correlations[41] exist? The answer is affirmative:

Theorem. *For every $c, \varepsilon, l > 0$ and long enough $n > l$, and $k < n$, the correlation given by the substring*

$$0^{\frac{k}{l}} 1^{\frac{n}{k(l-1)}}, \tag{5.10}$$

$n - k < \frac{n}{k(l-2)} - \varepsilon$ *is not spurious.*

Indeed, long enough c-random strings are Borel normal [61, p. 129]; in particular, in such a c-random string, the number of 0s is almost equal to the number of 1s. However, every c-random string of length n containing the correlation (5.10) cannot satisfy Borel normality, a contradiction.

The probability of a non-spurious correlation in an n-bit-string is less than

$$(n - k + 1) \times 2^{1-n/2} \longrightarrow_{n\to\infty} 0.$$

Here is an illustration. The best average rate of algorithmic compressibility is about 86.4% (i.e. a string x is compressed into Zip(x) whose length is about 86.4% of the length of x). The probability that a binary string x of length 2048 (relatively short) is reduced by 13.6% is smaller than

$$2^{-\frac{136}{1000}\cdot 2048} < 10^{-82},$$

a tiny number ($10^{82} \approx$ the number of hydrogen atoms in the universe), so with probability more than

$$1 - 10^{-82},$$

a database will be Zip-incompressible, so Zip-random. *All its correlations will be spurious!*

[41] A non-spurious correlation should not appear in any c-incompressible n-bit-string.

Chapter 6

A Formal Approach to Incompleteness and Halting

6.1 Axiomatic systems

The set of natural numbers $\mathbf{N} = \{0, 1, 2, 3, 4, \ldots\}$ can be defined axiomatically using the system Peano's axioms (1889). The language consists of a single unary function symbol S (short for "successor") and the axioms:

A1 There is a natural number 0.
A2 Every natural number n has a successor, denoted by Sn.
A3 There is no natural number whose successor is 0.
A4 Distinct natural numbers have distinct successors: if $n \neq m$, then $Sn \neq Sm$.
Induction If a property $P(n)$ depending on the natural numbers is true for 0 and for the successor of every natural number for which it is true, then $P(n)$ is true for all natural numbers.

The natural numbers can be obtained by iterating the successor function S: $1 = S0, 2 = SS0, 3 = SSS0, \ldots$. *Peano Arithmetic* (**PA**) is a more elaborated axiomatic system [366] "aiming" to prove all properties of natural numbers:[1] From Gödel's Incompleteness Theorem, it follows that **PA** *is sound, incomplete and undecidable.*

Propositional calculus studies how simple statements can be combined to form more complex statements and how the truth or falsity of complex statements is related to their components. The axiomatic system for the propositional calculus (**PropC**) [35] consists of

(1) a (finite or countably infinite) set A of elements called *propositional variables*,

[1] It was claimed that A. Wiles' proof of Fermat's Last Theorem is formalisable in **PA**.

(2) two logical connectives, negation ($\neg$) and implication ($\rightarrow$),
(3) the set L of *well-formed formulas*, shortly, *wff's*, defined inductively by the following rules:[2]

- $A \subseteq L$,
- if p, q are in L, then $\neg p$ and $p \rightarrow q$ are in L,
- nothing else is included in L,

(4) the *axioms*[3]: for all wff's p, q, r:

Axiom 1 $(p \rightarrow (q \rightarrow p))$,
Axiom 2 $((p \rightarrow (q \rightarrow r)) \rightarrow ((p \rightarrow q) \rightarrow (p \rightarrow r)))$,
Axiom 3 $((\neg q \rightarrow \neg p) \rightarrow (q \rightarrow p))$,

(5) the inference rule *modus ponens*: if p and $p \rightarrow q$, then q.

A wff q is *provable in* **PropC** (or is a *theorem of* **PropC**) if there exists a sequence of wff's $p_1, p_2, \ldots, p_n$ such that each p_i is an axiom or is obtained from p_s and p_t with $s, t < i$ by modus ponens, and $q = p_n$.

A *truth assignment* is a function that maps propositional variables to 1 (true) or 0 (false). A truth assignment a can be extended to wff's with the following rules:

(1) a satisfies the propositional variable p if $a(p) = 1$,
(2) a satisfies $\neg\varphi$ if a does not satisfy φ,
(3) a satisfies $\varphi \rightarrow \psi$ if it is not the case that a satisfies φ but not ψ.

A wff p is a *tautology* if $a(p) = 1$, for every truth assignment a. For example, the wffs $p \rightarrow p$ or $\neg\neg p \rightarrow p$ are tautologies. The following result was proven in 1921 by Post [323]:

Semantic Completeness Theorem for Propositional Calculus. *The set of theorems of* **PropC** *coincides with the set of tautologies.*

As a consequence, we can deduce the

Decidability Theorem for Propositional Calculus. *There is an algorithm deciding whether an arbitrary wff is a theorem of* **PropC**.

---

[2]Repeated applications of these rules permit the construction of complex formulas.

[3]Discovered by the Polish logician J. Łukasiewicz. Axiom 3 can be replaced by $(\neg(\neg p) \rightarrow p$. All three axiom can be replaced by a single one [282].

Consistency Theorem for Propositional Calculus. *For every wff p, it is impossible that both p and $\neg p$ are theorems of* **PropC**.

Predicate calculus (**PredC**) or first-order logic goes deeper into the structure of propositions and allows the use of sentences containing variables. For example, 'Socrates is a man' is a sentence in propositional calculus, but the expression 'there exists x such that x is Socrates, and x is a man' is not because it contains a variable (x) and a quantifier (there exists). The language of the axiomatic system of predicate calculus includes the propositional calculus language augmented with the following:

(1) quantifies, $\forall$ (for all) and $\exists$ (there exist),
(2) variables,
(3) non-logical symbols representing predicates (for example, a symbol for equality, $=$).

The wff's are constructed inductively in a similar manner, with two more rules:

(1) if P is a predicate of n variables and $v_1, v_2, \ldots, v_n$ are variables, then $P(v_1, v_2, \ldots, v_n)$ is a wff,
(2) if P is a wff and v is a variable, then $\forall v\ P(v)$ and $\exists v\ P(v)$ are wff's.

Like in the propositional calculus, but in a more elaborate form, an "interpretation", i.e. with a criterion for the truth of wff's, was developed for **PredC**. Accordingly, we have two classes of wff's again: the theorems, i.e. provable wff's, and the tautologies, i.e. the true wff's; see [35] for more details.

In 1929 Gödel [192, 193] proved

Semantic Completeness Theorem for Predicate Calculus. *The set of theorems of* **PredC** *coincides with the set of tautologies.*

As in propositional calculus, we have

Consistency Theorem for Predicate Calculus. *For every wff p, it is impossible that both p and $\neg p$ are theorems of* **PredC**.

So, anything proven for propositional calculus seems true also for the predicate calculus ... Not so! In 1936 Church [125] proved:[4]

Undecidability Theorem for Predicate Calculus. *It is undecidable whether a wff of the predicate calculus is provable (or, equivalently, true) in* **PredC**.

6.2 An automated proof of the Halting Theorem

Isabelle is a generic proof-assistant derived from the Higher Order Logic (HOL), a descendant of Logic for Computable Functions (LCF). LCF was developed in 1972, HOL became stable around 1988, and the development of Isabelle started in the 1990s [200]. Isabelle is based on a small core set of logical principles from which theories can be built. As such, the confidence with which we can claim any theorem proven in Isabelle to be true is the same confidence with which we can claim that the small core is true.

Isabelle provides a formal language to work in and a set of proof methods, which allows it to prove statements using logical rules, definitions, axioms, and already-proven statements. Proofs in Isabelle are written in natural deduction style. A structured proof language, Isar, is provided, which aims to make proofs more human-readable and greatly reduces the learning curve required to use Isabelle. In addition to the standard proof methods, Isabelle has a feature called *sledgehammer*, which calls external automated theorem provers to prove the current goal. Isabelle is developed jointly at the The University of Cambridge, Technische Universität München and Université Paris-Sud [432].

An Isabelle proof proceeds interactively. The user makes a claim and must then prove it. Isabelle's output (separate from the source code the user writes) gives information like the goals currently needing to be proven and whether any redundancy in proofs has been detected. The user can select a line of a completed proof and check the output to see what was being done at that point – this "hook" into the proof allows for an easier understanding of new proofs. Note that the Isabelle output is not included with the formal proof, which the user ends up with.

A detailed mechanised proof of the Halting Theorem, including proofs of correctness for all programs used, was given in [417] and discussed in [110].[5]

[4]For a historical account, see Section 7.2.

[5]The last article also contains a detailed analysis of the difference between correctness and termination in Isabelle.

The proof uses the Turing machine model of computation and follows, in broad strokes, the classical proof [147]. It assumes the existence of a Turing machine H such that given an encoding $\langle M, n\rangle$ of a Turing machine M and input n, H outputs 0 if M halts on n and 1 otherwise.

The following modification D of H is then constructed:

$$D\langle M\rangle = \begin{cases} \infty, & \text{if } M\langle M\rangle \text{ halts,} \\ 1, & \text{if } M\langle M\rangle \text{ does not halt,} \end{cases}$$

and a contradiction is reached by computing:

$$D\langle D\rangle = \begin{cases} \infty, & \text{if } D\langle D\rangle \text{ halts,} \\ 1, & \text{if } D\langle D\rangle \text{ does not halt.} \end{cases}$$

Any formalisation requires several aspects of the proof to be made explicit. For example, the modification D must be constructed, and the changes to H must be shown to be correct. Furthermore, explicit notions of halting and correctness must be defined. Since proofs are computer-checked, special care must be taken to implement halting.

The formalisation of Turing machines uses a two-way infinite single-tape Turing machine in which tape cells can be in one of two states – blank or occupied. The tape is represented by a pair (l, r) of lists, with l representing the cells to the left of the read/write head and r the cell being read and those to the right. Five actions are available; write blank, write occupied, move left, move right and do nothing. A Turing program is simply a list of pairs of actions and natural numbers representing states – the order of the pairs encodes which instruction maps to which state and input. Here is an example of program [417]:

$$\mathit{dither} := [\underbrace{\overbrace{(W_{Bk}, 1)}^{\text{read } Bk}, \overbrace{(R, 2)}^{\text{read } Oc}}_{\text{state 1 (start)}}, \underbrace{(L, 1), (L, 0)}_{\text{state 2}}]$$

The program begins in state 1. If it reads a blank cell, it writes it and goes to state 1. If it reads an occupied cell, it moves right and goes to state 2. In state 2, it moves left, returning to state one if it sees a blank cell and going to state 0 (the halting state) if it sees an occupied cell. Hence this program halts on a tape containing two occupied cells and loops indefinitely on any tape with fewer such cells.

Sequential composition of Turing programs is introduced to ease the construction of programs. Essentially, this modifies the programs by increasing the state numbers of any subsequent programs and changing the halting state to the start state of each next program. This composition is used to combine three Turing programs: a copy program (to copy a machine's code so it can read it), the supposed program to solve the Halting Problem (H above), and a program to loop infinitely in certain cases (the "dither" program). *Both the copying program and dither must be proved correct.* We will outline above how this proceeds for dither.

The correctness of a Turing program is established through the use of Hoare triples. The triple $\{P\}p\{Q\}$ indicates that the program p run on a tape satisfying P will result in a tape satisfying Q. We can also write $\{P\}p \uparrow$ to indicate that a p run on a tape satisfying P will never halt. Hence, the program dither should satisfy the triples:[6]

$$\{\lambda tp.\exists k.tp = (Bk^k, \langle 1\rangle)\}\ dither\ \{\lambda tp.\exists k.tp = (Bk^k, \langle 1\rangle)\}$$
$$\{\lambda tp.\exists k.tp = (Bk^k, \langle 0\rangle)\}\ dither\ \uparrow$$

The first statement can be established in Isabelle by calculation; provide a tape matching the first condition, run the dither program and see what happens. Due to the design of the implementation, this is straightforward and very easily automated. A proof of the second statement clearly cannot proceed in the same manner – *running dither on such a tape should result in an infinite loop* – a phenomenon which will affect any Isabelle simulation of the machine. Instead, the second statement can be established by induction on the number of steps performed, starting with the input tape.

Having established the correctness of copy and dither, one can prove the Halting Problem's undecidability, following the standard method. At this stage, a definition of the Halting Problem must be defined. The property of a Turing machine p halting on an input n is defined using Hoare triples as follows:[7]

$$halts\ p\ n := \{\lambda tp.tp = ([], \langle n\rangle)\}\ p\ \{\lambda tp.\exists k, m, l.tp = (Bk^k, \langle m\rangle @Bk^l)\}.$$

We then assume that a machine H exists, which solves the Halting

[6]In what follows, we use lambda abstraction, `https://wiki.haskell.org/Lambda_abstraction`. Lambda (λ-) calculus is a formal system introduced by A. Church for expressing computation based on function abstraction and application using variable binding and substitution. This model is universal as it can simulate any Turing machine.

[7]In Isabelle, the @ symbol indicates a concatenation of lists. Also, note that this definition of *halts* assumes functions with some number of inputs and a single output.

Problem. Formally within Isabelle, this is defined by the following Hoare triples:

$$halts\ M\ n \rightarrow \{\lambda tp.tp = ([B^k], \langle(\langle M\rangle, n)\rangle)\}\ H\ \{\lambda tp.\exists k.tp = (Bk^k, \langle 0\rangle)\}$$
$$\neg halts\ M\ n \rightarrow \{\lambda tp.tp = ([B^k], \langle(\langle M\rangle, n)\rangle)\}\ H\ \{\lambda tp.\exists k.tp = (Bk^k, \langle 1\rangle)\}.$$

Then we define the diagonalising Turing machine *contra* by

$$contra := copy; H; dither$$

where ; indicates the sequential composition of the programs. The contradiction is now reached through reasoning established from the proofs of correctness for copy and dither and the Hoare triple assumptions for H.

6.3 A unified presentation of Incompleteness Theorems based on the Halting Theorem

This section presents an abstract form of the First and Second Incompleteness Theorems and some consequences. The framework is justified by Gödel's view of a formal system as just a mechanical procedure for generating "provable formulas" [197]:

> ... due to A. M. Turing's work, a precise and unquestionably adequate definition of the general concept of formal system can now be given, the existence of undecidable arithmetical propositions and the non-demonstrability of the consistency of a system in the same system can now be proven rigorously for every consistent formal system containing a certain amount of finitary number theory.
>
> Turing's work gives an analysis of the concept of "mechanical procedure"(alias "algorithm" or "computation procedure" or "finite combinatorial procedure"). This concept is shown to be equivalent to that of a "Turing machine". ... *A formal system can simply be defined to be any mechanical procedure for producing formulas, called provable formulas.*[8]

We will work with an axiomatic theory having (CEN) and (CAR)[9] (shortly, *formal theory*) **T** "for almost all mathematics". An example is **ZFC**, Zermelo-Fraenkel set theory with the Axiom of Choice, the standard axiomatic set theory for mathematics. This means that virtually every

[8]Author's italics.

[9]The system contains a modicum of arithmetic of natural numbers; see Section 2.1.

mathematical proof, and, in particular, the proofs in this section, can be carried out in **T**.

A formula (sentence) S is provable in a formal theory **T** if there exists a proof π in **T** for S; in this case we write $\vdash S$. The formula $\neg S$ is the negation of S.

Consider a formal theory **T** using sentences over an alphabet Σ containing the symbol $\neg$. In **T** we fix two sets: P_1 is a non-empty computable set of sentences that do not start with $\{\neg\}$ and $P_2 = \{\neg S \mid S \in P_1\}$. We also define two sets of sentences in P_1 provable in **T**:

$$\mathrm{Prov}_1 = \{S \in P_1 \mid \vdash S\}, \mathrm{Prov}_2 = \{S \in P_1 \mid \vdash \neg S\}.$$

First Incompleteness Theorem. *Let **T** be a formal theory such that* Prov_1 *is not computable. Then, there exists a sentence in* P_1 *such that neither itself nor its negation is provable in **T**.*

From (CON), it follows that $\mathrm{Prov}_1 \cap \mathrm{Prov}_2 = \emptyset$. Assume by absurdity that for every $S \in P_1$, we have either $\vdash S$ or $\vdash \neg S$. Fix $S \in P_1$ and computably enumerate all proofs $\pi_i, i = 1, 2, \ldots$ in **T**; from the absurdity assumption, it follows that eventually either a proof π_i for S or a proof π_j for $\neg S$ will show up in the enumeration. In the first case $\vdash S$, so $S \in \mathrm{Prov}_1$; in the second case $\vdash \neg S$, so $S \in \mathrm{Prov}_2$, hence by consistency, $S \notin \mathrm{Prov}_1$. This proves that Prov_1 is computable, a contradiction.

We note that the First Incompleteness Theorem is syntactic: it says that *a formal theory such that* Prov_1 *is not computable does not satisfy* (SCO), *i.e. it is syntactic incomplete.*

A sentence that neither itself nor its negation is provable in **T** is called *undecidable* (in **T**).

In the context of the First Incompleteness Theorem, if S is undecidable in **T**, then the set $P_1 \setminus \{S\}$ is computable and the set $\mathrm{Prov}_1 \setminus \{S\}$ is not computable, hence the theorem can be applied again to $P_1 \setminus \{S\}$. We obtain:

Corollary. *In a formal theory **T** for which* Prov_1 *is not computable, there exist infinitely many undecidable sentences.*

The sentences $N(P, v)$ discussed in Section 2.2 can be formalised in a formal theory **T**, and they form a computable subset P_1. As a consequence of the undecidability of the Halting Problem, the set Prov_1 is computably enumerable but not computable, so we get Davis' Theorem [149]:

Syntactic version of Davis Incompleteness Theorem. *In every formal theory* $\mathbf{T}$ *here exist infinitely many undecidable sentences* $N(P, v)$.

We may still be able to prove that a particular sentence $N(P, v)$ is provable in $\mathbf{T}$ by a logical analysis of P's behaviour on v, and, indeed, there are infinitely many such cases.

Next, we recall the complexity measure K discussed in Section 2.5, which is used in Chaitin Incompleteness Theorem [113]. We replace the sentences $N(P, v)$ by '$K(v) > n$', and we note that the new set of sentences is also a computable subset P_1. Hence we obtain the following

Syntactic version of Chaitin Incompleteness Theorem. *In every formal theory,* $\mathbf{T}$ *here exist infinitely many undecidable sentences '*$K(v) > n$*'.*

Next, we present a form of semantic incompleteness in the same framework. We need a suitable formal definition of "true sentence" of $\mathbf{T}$ to this aim. The formulations of truth in the Semantic Completeness of the Propositional and Predicate Calculi are not powerful enough for the complexity of the formal systems aiming to be foundations for the whole mathematics. Following [279, p. 24] we can ask:

> ... how can we state, in Mathematics, that an assertion is true without demonstrating it (or taking it for hypothesis)? The interlocutor must then produce a proof convincing us of the "(unprovable) truth" of G.[10] The hypothesis of consistency, he/she points out, implies that G is unprovable (first theorem). And since G "asserts" that it is not provable ... then it is true. This reasoning based on the "meaning" of G is informal, vague and unwritten. But once formalised, it is a semantic version of the rigorous formal implication $\mathbf{PA} \vdash (Cons \rightarrow G)$ that constitutes the core of the second theorem.

In fact, matters are worse because of

Tarski Undefinability Theorem. *Arithmetical truth cannot be defined in Peano Arithmetic.*[11]

[10] This refers to Gödel's original proof [194].

[11] The result was proven in [388]. However, Gödel discovered the result in 1930 while proving his First Incompleteness Theorem [194]; see [304].

Still, a painless formalisation of truth suited for our discussion can be provided [71]. Motivated by the simple structure of sentences $N(P, v)$ and '$K(v) > n$', we introduce the following set:

By $True$, we denote a subset of $P_1 \cup P_2$ such that for every $S \in P_1, S \in True$ if and only if $\neg S \notin True$. The elements S in $True$ are called *true sentences* and are denoted by $\models S$. True sentences satisfy a semantic form of consistency:

> **Semantic Consistency Lemma.** *There is no $S \in P_1$ such that $\models S$ and $\models \neg S$.*

Fix a set $True$. We say that

- **T** is *Sound* for $True$ if for every sentence S in **T**, if $\vdash S$, then $\models S$.
- **T** is *Semantically Complete* for $True$ if for every sentence S in **T**, if $\models S$ then $\vdash S$.

> **Consistency Lemma.** *Every formal theory **T** which is sound for $True$ is consistent.*

Furthermore, if **T** is sound for $True$ and $Prov_1 \neq \emptyset$, then $P_1 \cap True \neq \emptyset$.

We are now in a position to state and prove

> **Semantic Incompleteness Theorem.** *Let **T** be a formal theory which is sound for a set $True$, and* Prov_1 *is not computable. Then, there exist in $True$ infinitely many undecidable sentences (in **T**).*

For the proof, we assume by hypothesis that **T** is sound for $True$, so by Consistency Lemma, **T** is consistent; as Prov_1 is not computable, the First Incompleteness Theorem applies, so we get a sentence $G \in P_1$ such that G and $\neg G$ are not provable in **T**. From Consistency Lemma one and only one of these two sentences is true. Finally, we use the Corollary to First Incompleteness Theorem to get infinitely many undecidable true sentences (in **T**).

As consequences, we obtain Davis [149] and Chaitin [113] theorems.

> **Davis Semantic Incompleteness Theorem.** *Let **T** be a formal theory which is sound for the set $True$ of all true sentences $N(P, v)$. Then there exist infinitely many true sentences $N(P, v)$ unprovable in **T**.*

Chaitin Semantic Incompleteness Theorem. *Let $\boldsymbol{T}$ be a formal theory which is sound for the set True of all true sentences '$K(v) > n$'. Then, there exist infinitely many true sentences '$K(v) > n$' unprovable in $\boldsymbol{T}$.*

Chaitin Theorem in Section 2.6 can be re-formulated as:

Chaitin Theorem. *Let $\boldsymbol{T}$ be a formal theory which is sound for the set True of all true sentences '$K(v) > n$'. Then, there exists a positive integer L such that for every v, the statement '$K(v) > L$' cannot be proven in $\boldsymbol{T}$.*

At a superficial level, one may think that the statement says that Chaitin Theorem is true for almost all true sentences '$K(v) > L$', but this is false: there exist infinitely many true statements '$K(v) > L$' provable in **T**.

Next, we present the Second Incompleteness Theorem, which deals with the provability of the consistency of a formal theory **T**.

First, we need a definition of "truth" in a formal theory **T** with a set *True*. We say that

- **T** is *Reciprocally Sound* for *True* if for every $S \notin True$ we have $\vdash \neg S$.

In other words, **T** is Reciprocally Sound when **T** proves the negation of every statement $S \notin True$: $\vdash \neg S$.

Second Incompleteness Theorem. *Let $\boldsymbol{T}$ be a consistent formal theory which is sound and reciprocally sound for a set True, and the set* Prov_1 *is not computable. Then, $\boldsymbol{T}$ cannot prove its consistency.*

Here is the proof. Assume that the computable set P_1 (which is infinite because there exist infinitely many undecidable sentences) is injectively coded in the form $\{S(x) \mid x \in \{0,1\}^*\}$, where $\{0,1\}^*$ is the set of all binary strings. For each $n \geq 1$ consider the set

$$M_n = \{S(x) \in P_1 \mid S(x) \in True, \mathrm{length}(x) \leq n\}.$$

By the Semantic Incompleteness Theorem, there exists a sentence $S(z) \in True \setminus Prov_1$, hence for every $n \geq \mathrm{length}(z)$, $S(z) \in M_n \setminus Prov_1$ and

$$1 \leq \#M_n \leq 2^{n+1} - 1. \tag{6.1}$$

Fix $n \geq \text{length}(z)$. If M_n has one element, then all the remaining $2^{n+1} - 2$ sentences $S(y)$ with $\text{length}(y) \leq n$ are not in M_n, hence not in $True$, so by reciprocal soundness, $\vdash \neg S(y)$. By computably enumerating the proofs $\pi_1, \pi_2, \ldots$ in **T** till all $2^{n+1} - 2$ proofs for these sentences show up, we get all $2^{n+1} - 2$ sentences $S(y)$ with $\text{length}(y) \leq n$ and $S(y) \notin True$;[12] the remaining unique sentence $S(z)$ is in $M_n \setminus Prov_1$. However, for every $\text{length}(y) \leq n, y \neq z$, we have $\vdash \neg S(y)$, and because **T** is rich enough, $\vdash S(z)$, in contradiction with the Semantic Incompleteness Theorem. Using the consistency of **T** and that **T** is rich enough, we deduce that **T** proves that M_n has at least two elements. The above reasoning can continue by assuming by contradiction that M_n has exactly two elements and getting, again, a contradiction with the Semantic Incompleteness Theorem. Step by step, we will reach the stage when **T** proves that M_n has more than $2^{n+1} - 1$ elements, a contradiction with (6.1).

As discussed in Section 2.2, we could see (and this is the author's opinion) that *the negative conclusion of the Semantic Incompleteness Theorem is a positive result*, even more than in the First Incompleteness Theorem, as we could not trust **T** to self-certify its consistency.

Corollary. *Let **T** be a formal theory which is sound for the set of all true sentences $N(P, v)$. Then **T** cannot prove its consistency.*

First, we note that the set $Prov_1$ is computably enumerable but not computable because of the Halting Theorem. The set $True = \{N(P, v) \mid P(v) \text{ does not stop}\}$ satisfies the required property; furthermore, **T** is reciprocally sound because if $N(P, v) \notin True$, then $P(v)$ stops, hence **T** can prove it by simply describing the running of P on v until it stops. Finally, given the consistency of **T** and the Consistency of Lemma, the conclusion follows from the Second Incompleteness Theorem.

As another corollary, we can prove the following result from [257]:

Kritchman-Raz Theorem. *Let **T** be a formal theory which is sound for the set of true sentences '$K(x) > n$'. Then **T** cannot prove its consistency.*

As in the proof of the above Corollary, Prov_1 is computably enumerable but not computable, the set $True = \{\text{'}K(x) > n\text{'} \mid K(x) > n\}$ satisfies the required property and **T** is Reciprocally Sound because if '$K(x) > n$'

[12]This process is analogous to the Decidability Criterion in Section 1.8.

$\notin True$, then there exists a program P and a string v such that $P(v) = x$ and length$(\langle P, v\rangle) \leq n$, hence **T** can prove it by giving P and v and describing the running of $P(v)$ until it stops and produces x. Again, the conclusion follows from the Consistency Lemma and the Second Incompleteness Theorem.

Finally, we present a proof by contradiction for the Chaitin Theorem and Chaitin Semantic Incompleteness Theorem (Complexity Variant) (see Section 2.6) to be contrasted with the derivation from the Semantic Incompleteness Theorem presented above. To this aim we construct the computable function p defined (on non-negative integers represented in binary) as follows:

> $p(i) = x$ if '$K(x) > s$' is the first theorem (in the order generated by the proofs of the formal system, a process guaranteed by (CEN)) of the form '$K(x) > s$' with $s \geq i$.

Because of soundness, $K(p(i)) > s \geq i$. In view of the Optimality Theorem in Section 2.5, there exists a constant c such that for all i:

$$i \leq s < K(p(i)) \leq K_p(p(i)) + c \leq \log_2(i) + c + c',$$

a contradiction.

Last we prove Chaitin Semantic Incompleteness Theorem (Complexity Variant). First, we recall that K is unbounded, that is, for every $n > 0$ there exists a string x such that '$K(x) > n$', hence there are infinitely many true statements of the form

$$\text{'}K(x) > m\text{'}$$

all expressible in the axiomatic theory. By Chaitin Theorem, there exists a constant c such that if '$K(x) > s$' is a theorem generated in the axiomatic theory, then $s \leq c$. Consequently, no statement in the infinite set $\{\text{'}K(x) > m\text{'} \mid K(x) > m > c\}$ is a theorem, despite all being true.

6.4 Unusually long proofs

Theorems can have short or long proofs. Long proofs are sometimes shortened. Does every theorem have a short proof? For formal theories, the answer is negative because each of them proves infinitely many theorems [309]:

Norwood Proposition. *For every formal theory* $\boldsymbol{T}$ *and positive integer* N*, there exist infinitely many theorems in* $\boldsymbol{T}$ *whose smallest proof lengths are bigger than* N.

The lengths of very long proofs have increased with time: 100 pages in 1900, 200 pages in 1950, and 500 pages in 2000. One of the longest mathematical proofs, measured by the number of published journal pages, is the classification of finite simple groups, written by about 100 authors, published chiefly between 1955 and 2004, which fills well over 10,000 pages [28]:

Finite Simple Group Theorem. *Every finite simple group is isomorphic to one of the following groups:*

(1) a member of one of the following three infinite classes:
  - *(a) the cyclic groups of prime order,*
  - *(b) the alternating groups of degree at least 5,*
  - *(c) the groups of Lie type,*

(2) one of the 26 "sporadic groups",

(3) the Tits group.

Many proofs (existence and uniqueness) for the "sporadic groups" have originally used computer calculations, but in time most of them have been replaced by shorter "hand-written" proofs.

Unusually long proofs use a combination of hand-written parts and computational results, which are problematic for some mathematicians and philosophers of mathematics. T. Tymoczko proposed the following three criteria to determine whether an argument is a mathematical proof:[13]

(1) *Convincingness* refers to the proof's ability to persuade a rational prover of its conclusion.
(2) *Surveyability* means the proof's accessibility for verification by human mathematicians.
(3) *Formalisability* denotes the proof's appealing to only logical relationships between concepts to substantiate its argument.

Computer-assisted proofs could be convincing and formalisable, but less or even not *surveyable* because of higher error risks and opacity; correctness is more difficult to assess and "understanding", a vital function of mathematics, could be concealed. For Teller [391] surveyability is a matter of

[13] In his critique [400] of Appel and Haken's computer-assisted proof of the Four Colour Theorem [27].

degree and reader-dependent, not a crisp property a proof has or does not have. A partial solution was used by Flyspeck project [430], which proved Kepler's conjecture:[14] the contentious parts of the proof were presented in a restricted formalism and a proof-assistant was used for surveying.

Can a human being read and understand a 1,000-page mathematical proof? Or a 5,000-page proof? Clearly, the answer is reader-dependent, but there is a limit. As computer-assisted proofs could produce unusually long proofs, it is natural to ask about their "meaning" for the working mathematician. To reflect on this question, we need to understand more about the lengths of proofs.

According to Spencer [376],

> Long proofs are anathema to mathematicians.

Gödel's seminal length-of-proof paper [195] was "re-discovered" after its English translation [198, pp. 396–399] and led to studies of theorems with long proofs [327, 328], and more generally, to Blum's (abstract) computational complexity theory [60].

But what is a "long proof"? Let us note that the original proof of a theorem tends to be unnecessarily long (and sometimes not entirely correct), but shorter and better proofs emerge in time. For example, Abel-Ruffini Theorem, stating the impossibility of finding a solution in radicals to polynomial equations of degree five or higher with arbitrary real coefficients, was initially 500 pages long (Ruffini's proof). Still, later, Abel obtained a mere 9-page proof.

A more interesting result than Norwood Proposition was proven by Hartmanis [216] for the class of formal systems whose theorems form a creative set. A computably enumerable set is *creative* if every other computably enumerable set can be one-one reduced to it.[15] All creative sets are recursively isomorphic [342].[16] The set of theorems of many interesting formal theories, including Peano Arithmetic and Zermelo-Fraenkel set theory, are creative. *Hartmanis proof-length* is the amount of tape used by Turing machines to enumerate the theorems of the formal theory. This measure

[14] Stating that no arrangement of equally sized spheres filling space has a greater average density than that of the cubic close packing (face-centred cubic) and hexagonal close packing arrangements.

[15] A set A is one-one reducible to a set B if there exists an injective computable function f such that for every string w, we have $w \in A$ if and only if $f(w) \in B$.

[16] A set A is recursively isomorphic to a set B if there exists a bijective computable function f from A to B.

is justified by the fact that for any formal theory one can design a Turing machine which, for any given sentence in the system, successively checks all possible proofs of increasing length until it finds a proof of the given sentence or never halts if the input is not provable in the formal theory.

> **Hartmanis Theorem.** *Fix a formal theory* $\boldsymbol{T}$ *whose set of theorems is creative, a Turing machine* M *that enumerates the theorems of* $\boldsymbol{T}$ *and Hartmanis proof-length with respect to* M*. Then, one can effectively find infinite subsets of "trivially true" theorems, which require as long proofs in* $\boldsymbol{T}$ *as the hardest theorems of* $\boldsymbol{T}$*.*

Here is a sketch of the argument. The proof consists in constructing a decidable infinite set of theorems S in $\mathbf{T}$, *TrivialTrue*, such that their shortest Hartmanis proof-length proofs in $\mathbf{T}$ grow faster than any computable function (of the length of the theorems to be proven). The theorems $S \in$ *TrivialTrue* are called "trivially true" because there is a decider for *TrivialTrue* which decides the question $S \in$ *TrivialTrue* with computably bounded space. The proofs in $\mathbf{T}$ of the theorems in *TrivialTrue* can be algorithmically generated by enumerating all proofs in $\mathbf{T}$ and selecting those whose corresponding theorems are in *TrivialTrue*.

Next, we will study the theorems with long proofs. To this aim we fix a formal axiomatic theory with negation $\mathbf{T}$ based on a finite alphabet. The formula $\neg S$ is the negation of S.

A formula (sentence) S is a *theorem of* $\mathbf{T}$, written, $\vdash^{\mathbf{T}} S$, if there exists a proof π in $\mathbf{T}$ for S, written, $\vdash^{\mathbf{T}}_{\pi} S$. The set of theorems in $\mathbf{T}$ will be denoted by $Thm(\mathbf{T}) = \{S \mid \vdash^{\mathbf{T}} S\}$.

Consider the following symbol-length measure: the *proof-length* of the proof π is the length of π (as a string on the finite alphabet of $\mathbf{T}$) and is denoted by $\mathrm{length}(\pi)$. The *minimum-length proof* of S is the shortest proof of S if S is provable in $\mathbf{T}$:

$$\pi(S) = \inf\{\mathrm{length}(\sigma) \mid \vdash^{\mathbf{T}}_{\sigma} S\}. \tag{6.2}$$

If there is more than one proof π satisfying the first condition in (6.2), then $\pi(S)$ is the first such proof in shortlex order.

First, we prove a stronger form of Spencer Theorem [376, 104]:

> **Long-Proof Theorem.** *Assume* $\boldsymbol{T}$ *is a computably enumerable, rich enough and consistent formal theory and* $f : \mathbf{N} \to \mathbf{N}$

a computable non-constant function. Then there exists an incomputable set of theorems I in $\boldsymbol{T}$ such that for every $S \in I$:

$$\text{length}(\pi(S)) \geq f(\text{length}(S)). \tag{6.3}$$

For a proof, assume by absurdity the existence of a computable non-constant function f such that for every theorem S in $\mathbf{T}$, we have

$$\text{length}(\pi(S)) < f(\text{length}(S)). \tag{6.4}$$

Under this assumption, we show that the set of theorems in $\mathbf{T}$ is computable, which contradicts Gödel's First Incompleteness Theorem [71] for $\mathbf{T}$. Indeed, the following algorithm decides membership in the set of theorems of $\mathbf{T}$. Given a formula S in $\mathbf{T}$:

(1) Calculate $f(\text{length}(S))$.
(2) Enumerate the finite set of proofs σ with $\text{length}(\sigma) < f(\text{length}(S))$.
(3) If for some proof σ, we have $\vdash^{\mathbf{T}}_{\sigma} S$, then return "yes" and stop; otherwise, return "no" and stop.

This algorithm clearly stops in finite time and returns the correct answer "yes". In case no proof σ proves S, then S is not a theorem of $\mathbf{T}$ because by (6.4), every theorem has a proof length $(\pi(S)) < f(\text{length}(S))$, hence the "no" answer is also correct.

Finally the set

$$\{S \in Thm(\mathbf{T}) \mid \text{length}(\pi(S)) \geq f(\text{length}(S))\} \tag{6.5}$$

is incomputable because otherwise the set $Thm(\mathbf{T})$ would be computable as the complement of the set (6.5) is computable, contradicting again Gödel's First Incompleteness Theorem [71] for $\mathbf{T}$.

Note the difference between the following two sets: a) $\{\sigma \mid \vdash^{\mathbf{T}}_{\sigma} S$, for some $S\}$, and b) $\{S \mid \vdash^{\mathbf{T}}_{\sigma} S$, for some $\sigma\}$. The first set is computable, but in the context of the Long-Proof Theorem, the second one is not.

The Long-Proof Theorem applies to Peano Arithmetic, Zermelo-Fraenkel set theory, the first-order theory of rational numbers with addition, multiplication and equality, and the first-order theory of groups. In contrast, Presburger arithmetic, the first-order theory of the natural numbers with equality and addition, the first-order theory of Euclidean geometry, and the first-order theory of Abelian groups are each decidable. Hence the Long-Proof Theorem does not work.

It remains, however, an interesting question whether the ratio of the length of proofs to the size of theorems is unbounded [309, p. 112]. The answer is:

Corollary. *Assume* $\boldsymbol{T}$ *is a computably enumerable, rich enough and consistent formal theory. Then, for every positive integer* N *there exists a theorem* S *in* $\boldsymbol{T}$ *such that* $\text{length}(\pi(S)) > N \times \text{length}(S)$.

For a proof, let $f(n) = n^2$ and use the Long-Proof Theorem to show the existence of an incomputable set of theorems I (depending on f) such that for each $S \in I$, $\text{length}(\pi(S)) \geq f(\text{length}(S)) = \text{length}(S)^2$. Giving a positive integer N we can choose $S \in I$ with $\text{length}(S) > N$ (because I is incomputable, hence infinite) so that $\text{length}(\pi(S)) > N \times \text{length}(S)$.

Corollary. *Assume* $\boldsymbol{T}$ *is a computably enumerable, rich enough and consistent formal theory. Then, for every positive integer* N*, the set.*

$$\{S \in Thm(\boldsymbol{T}) \mid \text{length}(\pi(S)) > N \times \text{length}(S)\} \qquad (6.6)$$

is incomputable.

The complement of the set (6.6) is computable, so by the Gödel's First Incompleteness Theorem [71] for **T**, the set (6.6) is computably enumerable and incomputable.

Recall that the sentence $N(P, v)$ says that the Turing machine P never halts on input v. The sentence $N(P, v)$ can be formalised in a sufficiently complex formal theory **T** like **PA** or **ZFC**. In such a **T**, we choose P_1 as the set of sentences $N(P, v)$. In the context of the First incompleteness Theorem the set Prov_1 is computably enumerable but not computable; in fact Prov_1 is creative. Hence, by Hartmanis Theorem, we get:

Corollary. *One can effectively construct infinite subsets of "trivially true" sentences* $N(P, v)$ *that require as long proofs as the hardest theorems of* $\boldsymbol{T}$.

We are now positioned to propose a definition of long/short proofs. Let $f : \mathbf{N} \to \mathbf{N}$ be a computable non-decreasing function. For example, $f(n) = n + \log n$. We say that a proof π for S is *f-short* if $\text{length}(\pi) \leq f(\text{length}(S))$; otherwise, π is *f-long*.

Below is an analogue of the Long-Proof Theorem for theorems with long statements and short proofs:

> **Theorem.** *Assume* $\boldsymbol{T}$ *is a computably enumerable, rich enough and consistent formal theory. Then, there exists an infinite computable set of theorems in* $\boldsymbol{T}$ *with* $n + \log n$*-short proofs.*

The proof is by construction: consider the theorem S_x = "2^{1x} is even", where x is a non-empty binary word. The proof π "As 2^{1x} is a positive power of 2, hence it is even" has length(π)+ constant < length(S_x) + log length(S_x), whenever length(x) is long enough. By varying x, we get a computable set whose elements S_x have the required property.

The results presented above show that formalising mathematics comes with price tags, which include unprovable statements and the existence of infinite sets of "trivially true" theorems that have very long proofs. They also disprove the view that long proof guarantees non-triviality/importance of a theorem and point to the importance of searching various formalisations and exploring new axioms [176].

6.5 Is Gödel incompleteness an exception?

Recall that a statement is independent with respect to an axiomatic system if itself and its negation are not provable in the system. For example, the Continuum Hypothesis is independent of **ZFC**, see Section 4.14.

Are independent statements featured in Gödel Incompleteness Theorem rather unusual, if not abnormal, statements? Indeed, this impression is recounted in [254, p. 399] by Solovay's "The feeling was that Gödel's theorem was of interest only to logicians" and Smoryński's "It is fashionable to deride Gödel's theorem as artificial, as dependent on a linguistic trick."

In this section, we show that the set of independent sentences in Gödel Incompleteness Theorem is "topological" large, a notion which will be defined in what follows.

All axiomatic systems in this section will satisfy (CEN), (CON) *and* (CAR).

> **Proposition.** *Let* $A \subseteq X^*$ *be an incomputable enumerable set expressible in an axiomatic system. Then, for some* $x \in A$, *the axiomatic system does not prove the (true) sentence "*$x \in A$*".*

Indeed, the set of $x \in A$ which the axiomatic system proves the sentence "$x \in A$" is computably enumerable.

Let the set $A \subseteq X^*$ be incomputable enumerable and expressible in the axiomatic system and let

$I(A) = \{u \mid u \in A, \text{ the axiomatic system does not prove "}u \in A\text{"}\}.$

Clearly, $I(A)$ is infinite, so it makes sense to answer the question: *How large is $I(A)$?* The main problem in addressing this question is to find a mathematical way of evaluating the size of $I(A)$ in such a way that degenerate, artificial or non-representative cases do not distort the answer.

Topology is a mathematical domain that studies spaces' invariant properties under continuous deformation. For example, a square can be deformed into a circle without breaking it, but an eight curve cannot. A square is topologically equivalent to a circle but different from an eight curve.

In what follows, we measure largeness in terms of topology showing that for many reasonable topologies, the set of these statements is *co-rare*, that is "large with respect to the topology" [63].

A closure operator on a set S is a function $cl : 2^S \to 2^S$ from the power set of S to itself satisfying the following conditions for every $X, Y \subseteq S$:

- $X \subseteq cl(X)$,
- if $X \subseteq Y$, then $cl(X) \subseteq cl(Y)$,
- $cl(cl(X)) = cl(X)$.

Every topological closure operator on a set gives rise to a topological space whose closed sets are precisely the closed sets. A set X is *closed* if $X = cl(X)$. The complement of a closed set is called *open.* An open neighbourhood N_x of a point x is an open set containing it. Vice-versa, by cl_τ, we denote the closure operator induced by the topology τ.

A partial order $\leq$ on X^* defines a topology $\tau_\leq$ as follows: For any $u \in X^*$, let $N_u^\leq = \{v \mid v \in X^*,\ u \leq v\}$. The closure operator $cl_{\tau_\leq}$ is given by

$$cl_{\tau_\leq}(A) = \{u \mid u \in X^*,\ \exists v \in A : u \leq v\},$$

for $A \subseteq X^*$. A set $A \subseteq X^*$ is *saturated* with respect to an equivalence relation if A is a union of equivalence classes.

We consider topologies τ on X^* having the following property:

> **(FIN)** There is a computable equivalence relation $\equiv$ on X^* such that, for every $x \in X^*$ and every open neighbourhood N_x of x, the set $\{y \mid y \in X^*, N_x \cap [y]_\equiv = \emptyset\}$ is finite.

The condition above excludes situations in which equivalence classes tend to form "clusters"; in such a case, clearly, no meaningful statement about the topological size of a set can be made.

The topologies induced by the shortlex and prefix orders are computable and satisfy the above condition. The set of computable orders satisfying the above requirement is invariant under computable bijections.

A set $A \subseteq X^*$ is said to be *rare* with respect to τ if, for every $x \in X^*$ and every open neighbourhood N_x of x, one has $N_x \not\subseteq cl_\tau(A)$. A set subset of X^* is *dense* if its closure equals X^*. It is *co-rare* if its complement is rare. Intuitively, the properties of being rare, dense, and co-rare describe an increasing scale for the sizes of subsets of X^* with respect to the topology τ. Thus a dense set is "larger" than a rare one and a co-rare set is "larger" than a dense set.

Theorem. [63] *Let $\equiv$ be a computable equivalence relation on X^* and let $A \subseteq X^*$ be a set which is expressible in an axiomatic system saturated by $\equiv$ and not computably enumerable. Then the set $I(A)$ is dense in every topology τ on X^*, which satisfies* (FIN) *with respect to $\equiv$.*

Theorem. [63] *Suppose that the topology τ is generated by a computable and length preserving partial order and satisfies* (FIN) *with respect to a computable equivalence relation $\equiv$. Let $A \subseteq X^*$ be a set expressible in an axiomatic system saturated by $\equiv$ and not computably enumerable. Then the set $I(A)$ is co-rare in τ.*

The above result applies to all important undecidable problems in the theory of computation, from the Halting Problem, the Totality Problem, the Emptiness Problem and the Post Correspondence Problem [67] to the Hamiltonian Problem in Hamiltonian mechanics [146], and to many others. The set of independent instances of any of these problems is co-rare for any topology generated by a computable and length-preserving partial order, hence *topological large*.

6.6 Every function is computable

First, "function" in the title of this section refers to a function from natural numbers to natural numbers.

Second, the sentence in the title cannot be correct. Why? Because of the Halting Theorem. Suspicious!

Recall the Peano Axioms and Peano arithmetic **PA** discussed in Section 6.1. The natural numbers $0, 1, 2, \ldots$ satisfy the Peano Axioms, so they are a *model* for **PA** called the *standard model*; after all, the axioms have been designed to "fit" the natural numbers. In general, a model for **PA** is a set endowed with a successor S satisfying Peano Axioms. It is not difficult to see that if $(\mathbf{N}, S, 0), (\mathbf{N}', S', 0')$ satisfy Peano Axioms, then they are isomorphic, i.e. there exists a bijection $\phi\colon \mathbf{N} \to \mathbf{N}'$ such that $\phi(0) = 0'$ and $\phi(S(x)) = S'(\phi(x))$, for all $x \in \mathbf{N}$. In other words, Peano Axioms characterise $\mathbf{N}$ up to an isomorphism.

Does this mean that Peano Axioms characterise only one model up to an isomorphism? No, because there are non-standard models for **PA**. These models contain extra numbers, called *non-standard* natural numbers, besides the standard ones! What is in the landscape of non-standard natural numbers? Non-standard natural numbers can be viewed as numbers bigger than every natural number; the successor function S is defined so that all these numbers satisfy Peano Axioms too. One way to construct a non-standard number is to take an undecidable statement G in **PA**; that is, one which is not provable nor disprovable in **PA**, see Section 2.2. If **PA** is consistent, then G holds in some models, while its negation holds in others. Like in physics, there exist different models or "possible worlds" of **PA**: every undecidable statement holds in some worlds, while its negation holds in other worlds.

D. Hamkins proved the following powerful result:

> **Hamkins Theorem.** [210] *There is a Turing machine T such that for every function f from the natural numbers to the natural numbers, there is a* **model** *of* **PA** *such that in* **this model**, *T on every standard natural n as input, halts and outputs $f(n)$.*

Take f to be your favourite incomputable function discussed in this book. Hamkins Theorem states that there is a model of **PA** such that in this model, a Turing machine computes $f(n)$ on every standard number as input.

Intuitively, there is a possible world in which your incomputable function becomes computable: shocking! The proof is beautiful but non-trivial. Following Baez [31] let's ask:

> But what's really going on? How can using a non-standard model make an incomputable function become computable for standard natural numbers? Shouldn't nonstandard models agree with the

> standard one on this issue? After all, the only difference is that they have extra non-standard numbers tacked on after all the standard ones! How can that make a Turing machine succeed in computing f on standard natural numbers?

Of course, the best answer is to read Hamkins's proof. Can we "grasp" the argument in an informal manner? Baez's suggestion [31] helps: read the end of the statement of Hamkins Theorem carefully:

> T on every standard natural n as input, halts and outputs $f(n)$.

The Turing machine T halts on every standard natural n means that for every standard natural n, there exists a natural number N_n (which depends on n) such that T halts on n after N_n steps. Is N_n a standard natural number? We don't know: we only know that N_n is a natural number in the **model** of **PA**, so it can well be a non-standard one. Indeed, T on n stops on N_n steps, which can take "longer than any standard number of steps".

The proof shows that the Turing machine T computes different functions in different models of **PA**. How can it work? The Turing machine T does not halt in a standard time N in different models, so N has to be non-standard. This line of reasoning leads to other strange possibilities: a) the standard natural numbers could be more subjective than we like to think, and b) maybe, as E. Nelson [307] suggested, the finite set is an elusive concept and not the same for all mathematicians.

Chapter 7

Historical and Philosophical Notes

7.1 Computability forerunners

G. W. Leibniz was a prominent German polymath and philosopher, one of the Enlightenment's most important logicians, mathematicians and natural philosophers.

Fig. 7.1: G. W. (von) Leibniz[1]

The differential and integral calculus, the mathematical theory of continuous change, is probably his most notable accomplishment.[2] Calculus

[1]Source: In Wikipedia. `en.wikipedia.org/wiki/File:Christoph_Bernhard_Francke_-_Bildnis_des_Philosophen_Leibniz_(ca._1695).jpg`.

[2]The paternity of this discovery was the subject of a bitter dispute with I. Newton, who developed independently similar ideas but didn't publish his results. Eventually, both have been credited for independently inventing and developing calculus, a theory widely used in science, engineering, economics, and beyond.

does for continuous change what geometry does for shape and algebra for generalisations of arithmetic operations.

Leibniz is also famous for another immense discovery that today is taken for granted: a system of calculus based on the binary numeral system (using the bits $0, 1$), which evolved in what we call today algorithmic thinking. He didn't stop with the language and the system but went on to construct in 1671 the first mechanical calculator that could perform all four arithmetic operations, the "Stepped Reckoner"[3], dreamt of a universal machine which, by formal calculation, could determine the truth of all mathematical statements [37, p. 82]. This path led him to state the slogan "Calculemus!" ("Let us calculate!").

For Leibniz, calculation should be the key to settling all human conflicts and disagreements in mathematics, philosophy and life.

Fig. 7.2: Replica of Leibniz's stepped reckoner in the Deutsches Museum[4]

Leibniz noted that the Universe is highly diverse and complex, but it appears to be ruled by very simple fundamental laws, later convincingly illustrated by Wolfram [415]. This type of duality raises fundamental questions about Science and Nature, some of which were posed already by Leibniz himself [119]: Is there any difference between a Universe without laws of Nature and one described by laws? How can we tell the difference? For

[3]The picture in Figure 7.2 is reproduced with permission by User: Kolossos - recorded by me in Technische Sammlungen der Stadt Dresden (with photo permission), CC. BY-SA 3.0, commons.wikimedia.org/w/index.php?curid=925505.

[4]Source: In Wikipedia. en.wikipedia.org/wiki/File:Leibnitzrechenmaschine.jpg.

many years the answers were pretty aligned with the doctrine generated by the enormous success of the scientific method, but new results and arguments are challenging this view [66, 345, 100].

Fig. 7.3: C. Babbage[5]

The English mathematician, philosopher, inventor and mechanical engineer C. Babbage invented the first mechanical computer, the "Difference Engine". Eventually, he constructed the "Analytical Engine" in 1837, the first programmable mechanical general-purpose computer. According to the "Babbage Thesis" proposed by Gandy [189, pp. 55–111], "the whole of the development and operations of analysis are now capable of being executed by machinery"; furthermore, Gandy noted that when thought of as a register machine, the "Analytical Engine" is equivalent to a universal Turing machine.

A second trend leading to computability theory emerged from Dedekind [155] work on the definition of basic arithmetic functions (addition, multiplication, exponentiation) and Peano [314, 315] axiomatic definition of natural numbers, specifically his 5th axiom which introduces the mathematical induction. These tools were used by Gödel [194], Pèter [317] and Kleene [247] to introduce the definition by (scheme of) primitive recursion, an essential ingredient in the construction of the class of primitive

[5]Source: By Antoine Claudet – National Portrait Gallery, Public Domain, `commons.wikimedia.org/w/index.php?curid=48042816`.

recursive functions. This new class of functions of natural numbers played a crucial role in the foundations of mathematics, specifically in the work related to Hilbert's school of mathematical logic. The class proved to be strictly included in the class of recursive functions: motivated by a Hilbert metamathematical problem [222], examples of (general) recursive functions that are not primitive recursive have been constructed independently by his PhD students G. Sudan [383] and W. Ackermann [11]; see more in [65].

Fig. 7.4: Difference Engine, Computer Museum, University of Auckland[6]

According to Soare [368]

> These two trends of recursion and computability were brought together in the 1930s by Gödel, Church, Kleene, Turing, and others, partly in response to questions raised earlier by Hilbert.

7.2 Entscheidungsproblem and Turing

The Entscheidungsproblem (German, "decision problem") goes back (again) to Leibniz. More precisely, the Entscheidungsproblem, "the chief problem of mathematical logic", posed by D. Hilbert and W. Ackermann in 1928 [223], asked for an algorithm that takes as input an arbitrary statement of the first-order predicate calculus **PredC** and answers "Yes" or "No" according to whether the statement is provable or not from the axioms of **PredC** (equivalently, it is universally valid, i.e. valid in every structure satisfying the axioms of **PredC**).

[6]Source: From `www.cs.auckland.ac.nz/historydisplays/SecondFloor/CharlesBabbage/BabbageImages/ComputerMuseumDemo.jpg`.

In 1934, Turing finished with first-class honours in mathematics from King's College, Cambridge. Toward the end of that year, he began attending M. Newman's[7] *Foundations of Mathematics* course where he learnt about the Entscheidungsproblem and decided to remain in King's College to work on it.

At the time, the Entscheidungsproblem had been solved in the affirmative for various special classes of formulas (for example, propositional logic and the more interesting monadic first-order logic, that is, logic with only one-place predicates). Still, Turing became convinced that the answer to the general problem must be negative. Before going further, it is important to note the *crucial asymmetry between a positive and a negative answer.* A positive answer can be obtained by constructing an algorithm and proving it solves the problem. To understand such a proof, we need only to recognise an algorithm when we see it. A negative answer has to demonstrate that no such algorithm exists. Since there are infinitely many, we cannot prove this by an exhaustive search through all possible algorithms. Hence, to get a negative proof, we need some way of proving facts about *all algorithms*, so we need

(1) a mathematical characterisation of the infinite class of algorithms and
(2) a proof that no algorithm in the class solves the problem.

Working on such a characterisation, by mid-April 1936, Turing invented his "automatic machine" (a-machine) known today (after Church [56]) as the Turing machine, a fundamental notion in computability theory. How do we know that the proposed model is "adequate", i.e. it describes "correctly/faithfully" the informal notion of the algorithm? This question cannot typically be answered with a mathematical provable fact, but substantiated and then postulated as a thesis or adopted as a definition[8]. The adequacy of the Turing machine as a model for an algorithm is given by the Church-Turing Thesis [139]; see Section 1.12. According to [175, p.146]

> After some initial scepticism, Newman became convinced of Turing's analysis and encouraged its publication. Neither Newman nor Turing were aware at that point that there were already two other proposals under serious consideration for analyzing the general concept of effective computability: one by Kurt Goödel, building

[7]M. Newman, 1897–1984; not to be confused with J. Newman, co-author with E. Nagel of the famous book *Gödel's Proof* [305].

[8]Think about the mathematical definitions of a function and continuous function.

> on an idea of Jacques Herbrand, called general recursiveness, and the other by Church, called definability in the lambda λ-calculus. Church and his student Stephen C. Kleene proved the equivalence of these two notions, and what has come to be called 'Church's Thesis' was that general recursiveness constitutes an analysis of the concept of effective computability. This claim was stated in the paper [125], in which *various mathematical and logical problems were shown to be effectively undecidable.*[9] That was followed by [126], in which the general Entscheidungsproblem was answered in the negative, submitted 15 April 1936 just about the same time as Turing was preparing his paper for publication. When Newman and Turing received news about this work a month later,[10] the first reaction was a great disappointment, but then it was agreed that Turing's analysis was sufficiently different to warrant publication. Moreover, after that was submitted but before it appeared, Turing was able to add an appendix in which he proved the equivalence of computability by his machines with lambda-definability.

For all practical purposes, Turing's work was 'killed' by Church's papers [125, 126]: there is no second prize in mathematics. But there are exceptions; in this case, the departure from the norm was beneficial. When Newman realised the importance of Turing's proof and his different approach, he communicated with Church and the London Mathematical Society. Finally, he succeeded in publishing Turing's paper [397] at the end of 1936 (with a correction one year later). Newman suggested that Church invite Turing to work with him in Princeton, where Gödel was also working. And indeed, Turing went and got his PhD with Church in 1938.

7.3 The Halting Problem: a false version

According to Strachey [381]

> A well-known piece of folk-lore among programmers holds that it is impossible to write a program which can examine any other program and tell, in every case, if it will terminate or get into a closed loop when it is run.

The reader knows from Section 5.3 that this "simplified" version of the

[9]Author's Italics.
[10]More precisely a reprint of Church's paper. (Author's note)

Halting Problem, which still circulates among some beginner students and programmers, *is not impossible*. The quotation is interesting because of its continuation:

> I have never actually seen a proof of this in print, and though Alan Turing once gave me a verbal proof (in a railway carriage on the way to a Conference at the NPL in 1953), I unfortunately and promptly forgot the details. This left me uneasy feeling that the proof must be long or complicated, but it is so short and simple that it may interest casual readers. The version below uses CPL, but not in any essential way.

7.4 ChatGPT-4 on the Halting Problem

On 30 April 2023, I asked Chat-GPT-4 to explain what is the Halting Problem and to give a proof of the Halting Theorem and the answers were:[11]

> The Halting Problem is a famous problem in computer science that asks whether there exists an algorithm that can determine whether a given program will halt (terminate) or run forever (enter an infinite loop) on all possible inputs. In other words, it asks whether there exists a program that can predict with certainty whether another program will eventually stop running or not.
>
> The problem was first formulated by Alan Turing in 1936, as part of his work on the foundations of computer science. Turing proved that no such algorithm can exist, using a technique known as diagonalization. In other words, there is no general method that can always determine whether an arbitrary program will halt or not, regardless of how clever or sophisticated the algorithm is.
>
> The implications of the Halting Problem are profound, as it means that there are fundamental limits to what computers can do. It also has practical applications in fields such as program verification, where it is often useful to know whether a program is guaranteed to halt, and in cryptography, where the difficulty of solving certain problems is based on the assumption that no efficient algorithm exists for solving the Halting Problem.

[11] `chat.openai.com/c/83770082-c6c1-4a52-9423-6d868ef0099c`, `chat.openai.com/c/4122fbcb-0b09-42af-a71e-b91659ad5cf3`. This web link is author-specific: a copy can be obtained from the author.

The Halting Problem Theorem is a fundamental result in the field of computer science, which states that it is impossible to write a program that can determine whether another program will eventually halt or run forever. The proof of this theorem is based on a clever argument by contradiction.

Suppose, for the sake of contradiction, that there exists a program H that can solve the halting problem, i.e. given any program P and input I, H(P, I) returns "halts" if P halts on input I, and "runs forever" otherwise.

We can use this program H to construct a new program G, as follows:

python Copy code

```python
def G(P):
    if H(P, P) == "runs forever":
        while True:
            pass
    else:
        return
```

Fig. 7.5: Python code

Now let's consider what happens when we run G(G). If G(G) halts, then by definition of G, we must have H(G, G) = "halts". But if H(G, G) = "halts", then by the definition of G, we must have entered the "else" branch of the if statement, which means that G(G) must have returned without looping forever. This is a contradiction, so we can conclude that G(G) must run forever.

On the other hand, if G(G) runs forever, then by definition of G, we must have H(G, G) = "runs forever". But if H(G, G) = "runs forever", then by the definition of G, we must have entered the "if" branch of the if statement, which means that G(G) is looping forever. This is also a contradiction, so we can conclude that G(G) must halt.

Therefore, we have arrived at a contradiction, and we can conclude that our initial assumption that there exists a program H that can solve the halting problem must be false. Hence, the Halting Problem Theorem is proven.

Is the answer correct? The following paragraphs include the first two errors (marked in bold):

> The Halting Problem is a famous problem in computer science that asks whether there exists an algorithm that can determine whether a given program will halt (terminate) or **run forever (enter an infinite loop) on all possible inputs.** In other words, it asks whether there exists a program that can predict with certainty whether another program will eventually stop running or not.
> **The problem was first formulated by Alan Turing in 1936, as part of his work on the foundations of computer science. Turing proved that no such algorithm can exist, using a technique known as diagonalization.** In other words, there is no general method that can always determine whether an arbitrary program will halt or not, regardless of how clever or sophisticated the algorithm is.

The first error is clear. The second one will be discussed in the following section. Are there more errors?

7.5 More about the origin of the Halting Problem

The Halting Problem is not the first example of an undecidable problem: Hilbert's Entscheidungsproblem [223] – the problem of deciding for an arbitrary statement of first-order logic whether it is valid or not[12] – was the first stated and proved to be an undecidable problem. The first proof of this major result was due to A. Church who published it in April 1936 [126]; Turing's independent proof published in [397] used his (later-called) Turing machines and, as we discussed in Section 7.2, there is a strong connection between Entscheidungsproblem and the Halting Problem.

As mentioned in Chapter 1, the Halting Problem was named and first proved by M. Davis [147, pp. 70–71]. However, the formulation and undecidability proof of the Halting Problem is usually wrongly referred to Turing's seminal papers [397].[13] In contrast, more recent books and articles like [43, 154] cite the origin of the problem correctly. Lucas [281]

[12] The corresponding decision problem for the propositional logic is easily solved with truth-tables.

[13] Another instance of Stigler's law of eponymy, stating that no scientific discovery is named after its original discoverer [434]. For a list of misnamed theorems, see `https://en.wikipedia.org/wiki/List_of_misnamed_theorems`.

presents a comprehensive analysis of the authorship of the definition and proof of the unsolvability of the Halting Problem. Following this article, there are three arguments in favour of Davis' paternity:

(1) in contrast with Church's paper [126], Turing did not introduce the stopping state in [397],
(2) Strachey's second quotation in Section 7.3,
(3) the decidability problems Turing considers are different from the Halting Problem.

The first argument is not surprising because Turing's main interest was "ostensibly the computable numbers" [397, p. 230], that is, real numbers, like π, whose infinite sequence of digits can be computed by a Turing machine which *never stops*.

Strachey's second quotation suggests that Turing's results in [397] have not been considered by both Turing and Strachey as providing proof of the undecidability of the Halting Problem. Furthermore, the decidability problems studied by Turing in [397] consider sequences computable by circle-free Turing machines; by definition, these machines *never stop*. The very definition of a circle-free Turing machine is not clear, as it was pointed out in the Appendix of Post's paper reprinted in [148, p. 300]; for more technical issues in [397], see [435].

> Davis thinks it likely that he first used the term 'halting problem' in a series of lectures that he gave at the Control Systems Laboratory at the University of Illinois in 1952 (letter from Davis to Copeland, 12 Dec. 2001).

Finally, Davis' notion of computation [147] required that the Turing machine halts, which naturally led Davis to formulate the Halting Problem and proved its undecidability. However, the last sentence of his book [147, Chapter 5, Section 2] reads:

> It might also be mentioned that the unsolvability of essentially these problems was first obtained by Turing.

This unfortunate comment may have contributed to the wrong credit of the formulation and undecidability proof of the Halting Problem, an error which persisted for more than 65 years; see also [152].[14] A nice updated

[14] This is the review M. Davis referred to in his letter below.

timeline for the history of the Halting Problem can be found in [431].

During my work on this book, I exchanged with M. Davis the following two emails:

```
   15 August 2021

Dear Martin,

I am writing a book "To Halt or Not To Halt. That Is the
Question" and not surprisingly the origin of the Halting
Problem is discussed.

Lately a few papers dealing with this issue have been
published, including the attached one.
While it is well known and documented that you have named
the problem and proved it's undecidability, it is puzzling
that for a very long time Turing was credited with them.
Your silence on this matter, as far as I was able to
check, has probably contributed. Lucas also thinks that
your statement

"It might also be mentioned that the unsolvability of
essentially these problems was first obtained by Turing"
could have also played a role.

I would be grateful for any comment you may have on
this topic.

With thanks and best regards,
Cris

PS Incidentally, a long time ago I have studied the paternity
of the first example of a recursive function which
is not primitive recursive and the conclusion was that G.
Sudan (1926) and W. Ackermann (1927) should share it. Sudan
refused to discuss it.
```

```
16 August 2021

Dear Chris,

By coincidence I am reviewing Lucas's paper for Math
Reviews.

Also I participated via Zoom as a commentator on a
talk Lucas recently gave.

I'll send you my review when it is finished. Right
now my main computer seems to have died, and I need
to do something about that.

Martin
```

The review arrived on 10 October 2021.

7.6 Proof-assistants

The statement [32]

> This statement has no proof in **PA** that contains fewer than N symbols.

can be formulated in **PA** (using Gödel's method [194]) but cannot be proved with less than N symbols if **PA** is consistent. If we take an integer N larger than the number of particles of ordinary matter in the Universe, crudely estimated to 10^{80}, this proof cannot be written down even if one could write one symbol on each particle.

From an arithmetical point of view, the above statement is not particularly interesting. We have discussed unusually long proofs in Section 6.3. Still, are relevant examples of theorems with long proofs an exception? The answer is negative: see the examples of interesting theorems with long proofs in [433].

A seemingly naive question is: Can brute-force-proof-search be improved to become a helpful tool? The answer is related, at least in part, to the problem of "automating" mathematics, an example of which was discussed in Section 6.2.

Fix a formal theory for a part of mathematics, **A**. There are at least three interpretations of "automating **A**":

a) We can write an algorithm that decides whether an arbitrary statement in **A** can be proven or disproven in **A**.
b) We can write an algorithm that finds proofs for all provable statements in **A**.
c) An economically-viable algorithm can perform the human activity of proving theorems in **A**.

The alternative a) is valid for some **A** (like the propositional calculus) but false for more complex theories, like **PA** or **ZFC** for the whole mathematics, because of the Halting Theorem. The weaker interpretation b) is true because it does *not* require an algorithm to decide the provability status within **A** of an arbitrary statement. Indeed, a brute-force algorithm can find a proof for every statement which is *provable* in **A**. Such an algorithm is highly inefficient and impractical when proofs are very long. However, this is *not* a limitation affecting the working mathematician because humans cannot read, even less understand, "too long" mathematical statements.

Brute-force-proof-searches show that b) is possible, which is one reason for developing proof-assistants.

From b), we naturally arrive at c), which can be discussed from three points of view: computational complexity, economics, and epistemology.

Following Gödel [191], if **P**=**NP**, then there is a polynomial-time algorithm that given a first-order statement and a positive integer n (in unary), decides whether the statement has size n proof in **ZFC**. It may seem that under this hypothesis, there is no computational complexity obstacle to answering the question c) affirmatively. However, this is not true because the distinction between **P** and **NP** is mathematically, but *not* practically, meaningful: **P**=**NP** only implies that problems that can be verified in polynomial-time can also be solved in polynomial-time; however, polynomial-time algorithms are not necessarily practical [150]. A quadratic time algorithm can check very long proofs; in fact, proofs with longer than any proof a human can write. Hence, if the algorithm does not find an answer, then the statement is practically/humanly impossible to prove.

Is it enough to know that something follows from some axioms and rules of inference, or is a proof something that provides deeper insight? Understanding is crucial in mathematics, so what kind of statements could be "humanly interesting"? What is the meaning of such a statement? Is a practical prover epistemologically viable, too? Fortunately, proof-assistants are not ordinary algorithms working "in their world", but algorithms used in human-machine cooperation, in which *understanding* is essential and achievable [64, 101, 395].

Incompleteness can be proved by reduction to the Halting Theorem, Section 2.2. Can the approximate solutions to the Halting Problem – as the one discussed in Section 1.16 – give more information?

7.7 Why and what do we compute?

The questions in the title of this section have been posed and briefly discussed by Abramsky in [10].

The first question asks why we compute, in mind or with specifically designed machines. The most straightforward answer seems to be to gain or increase theoretical or practical knowledge and information for understanding or changing the subject or an object.

This answer raises the question of *possibility*: can computation produce or increase knowledge and information? As knowledge combines information, experience, and insight, we can restrict the question to information. But what is information? Intuitively, information is data that is processed, structured, and presented in a given context to make it meaningful and valuable. Despite having more than one theory of information, starting with Shannon's theory [355], algorithmic information theory [115], and many others [374], one could argue that information cannot be meaningfully captured in one general and compressive model. Accordingly, we will attempt to answer the question from only two points of view: mathematical and physical.

The data processing inequality is a fundamental mathematical result in Shannon's information theory that states that the information content cannot be increased via a local physical operation. Informally, *post-processing cannot increase information*, Theorem 2.8.1 in [143, p. 32]. A quantum version of the result [36] can be stated as follows: *one cannot increase the information content of a quantum system by acting on it with a local physical operation.* These information-theoretic results show that computation cannot increase information.

Another argument in favour of this conclusion comes from the Second Law of Thermodynamics, which states that the entropy of an isolated system – defined as the log of the number of microstates in the system – is non-decreasing. A formal justification of this claim comes from [297];

a short informal argument is: "What may start in an orderly state will – according to the Second Law – inexorably 'degrade' to a 'randomised' state" [411].

The scandal of deduction refers to the fact that conclusions based on logical deduction are "logically contained" in the premises, so they do not add anything new.[15] In more detail,

(1) The conclusion is not novel if the deductive argument is valid.
(2) If the conclusion of an argument is not novel, the argument is not useful.

Hence:

(3) If a deductive argument is valid, it is not useful.

But (3) is unacceptable: yet another argument that computation cannot increase information.

Since the reasoning seems valid, to avoid (3), we must reject at least one premise. So, should we reject (1) or (2)? This puzzle is usually known as the "scandal of deduction".

Analytic philosophers have assumed premise (2) and rejected (1). According to [167], Aristotle would have denied (2) because he assumed deductions could be useful even without presenting novel conclusions.[16]

Are these arguments as solid as they appear? The data processing inequality seems the strongest demonstration. The second law of thermodynamics could be partially bypassed by considering sub-systems of the isolated (total) system whose entropy cannot decrease. For example, a sub-system can decrease entropy by consuming energy from its environment [10]. Agents and their interactions are essential for studying information flow and can increase computation in both senses (e.g. factoring and multiplying positive integers). Current theories of information reflect poorly these phenomena. The scandal of the deduction is the most vulnerable argument. One of the issues is that if an agent knows a proposition, then it knows all its logical consequences, a practically false deduction: We compute to gain the information we did not have.

This brings in the question: What do we compute?

[15]What do we learn when we apply logic? According to the theory of classical semantic information [34], the answer is "nothing objective".
[16]Of course, this does not solve the puzzle.

Computing mathematical functions is an integral part of the massive computation running in the world. However, in much of contemporary computing, the purpose of the computing system is to exhibit specific behaviour. For example, Internet computing includes buying books from Amazon, booking accommodation with Airbnb, booking air tickets, etc. and computing mathematical functions for blockchain and cryptocurrencies [9]. Computing on the Internet is a combination of formal computation with human interaction.

Quantum and AI computing have recently become a huge part of our computing. Quantum computing is used for problems as diverse as logistics, portfolio optimisation, drug discovery, materials sciences, scheduling, fault detection, traffic congestion, and supply chain management. AI computing is used almost everywhere: for example, American Express uses AI computing to detect fraud in billions of annual credit card transactions, and doctors use it to find tiny anomalies in mountains of medical images. These two forms of computing work better in hybrid forms, for example, combining classical and quantum algorithms.

This gives rise to the question: Which new forms of computing will be developed?

7.8 A brief history of undecidability

Contemporary usage recognises two distinct meanings associated with the term "undecidable". The first meaning originates in Gödel's Incompleteness Theorems and refers to a statement that cannot be proven or disproven within a given formal system for some properties.

Cantor's Continuum Hypothesis (1978) states that:

> There is no set whose cardinality is strictly between the naturals and the real numbers.

He believed the Continuum Hypothesis was true and tried vainly to prove it. The combined work of Gödel [196] and P. Cohen [131] led to the proof that Cantor's Continuum Hypothesis is independent of Zermelo-Fraenkel (**ZF**) set theory. This means that either adding the Continuum Hypothesis or its negation as an axiom to (**ZF**) set theory, the resulting theory is consistent if and only if (**ZF**) is consistent.

The second meaning, used in computability theory, applies to decision problems rather than statements. Decision problems consist of countably infinite questions, each requiring a yes or no answer. A decision problem is undecidable if no computable function can correctly answer every question in the problem set.

The connection between these two meanings lies in the fact that if a decision problem is undecidable (in the sense of computability theory), then there is no consistent, effective formal system that can prove either "the answer to Q is yes" or "the answer to Q is no" for every question Q in the problem.

The undecidable statements produced by Gödel's Incompleteness Theorems are *relative* to the system. Undecidability implies that the formal system under consideration cannot prove the truth or falsity of the statement. This raises an interesting question: are there mathematical statements that are fundamentally undecidable, meaning they cannot be determined as true or false by any "reasonable" formal system? If it exists, such a statement would be called "absolute undecidable". From the beginning, Gödel noted that his original Incompleteness Theorems did not provide examples of absolute undecidability, thus not undermining Hilbert's belief that every precisely formulated mathematical question has a definite and discoverable answer. However, in his subsequent work on set theory, Gödel discovered what he initially considered a plausible candidate for an absolutely undecidable statement. He even hoped that it could be proven as such. Eventually, he shifted to the opposite stance, hoping that a generalised completeness theorem could exist, implying the absence of any absolutely undecidable statements.

Let us consider the following scenario. Let $\text{undec}(x)$ be a predicate in **ZFC** that says that "x is a statement that is neither provable nor refutable in **ZFC**". As this predicate is expressible in **ZFC**, we can keep iterating this "undec" operator. Then there exists a true statement S such that $\text{undec}(S)$ is true, $\text{undec}(\text{undec}(S))$ is true, $\text{undec}(\text{undec}(\text{undec}(S)))$ is true, and so on. Indeed, we can take S to be a true and undecidable statement in **ZFC**. If **ZFC** is consistent, we can never prove any statement of the form $\text{undec}(x)$ because that would prove the consistency of **ZFC**, contradicting the Second Incompleteness Theorem. Therefore each true statement $\text{undec}(x)$ implies the truth of the next one.

The existence of so-called "absolutely undecidable" statements, whose truth value can never be known or is ill-defined, remains controversial among various mathematical and philosophical schools. Gödel accepted Hilbert's rejection of the existence of absolutely unsolvable problems because otherwise,

> it would mean that human reason is utterly irrational by asking questions it cannot answer, while asserting emphatically that only reason can answer them ...

Gödel [199] related this question to the distinction between objective and subjective mathematics. Objective mathematics consists of the body of mathematical propositions, constructive or not, which are true in an absolute sense. **PA** and **ZFC** are parts of it. Subjective mathematics consists of all mathematical truths *humanly demonstrable* (or *provable* or *knowable*), in a constructive manner or not.

The question is:

> Does objective mathematics coincide with subjective mathematics?

Gödel's answer (in the Gibbs lecture "Some Basic Theorems on the Foundations of Mathematics and their Implications" [199, p. 310]) is based on his Incompleteness Theorem:

> So the following disjunctive conclusion is inevitable: Either mathematics is incompletable in this sense, that its evident axioms can never be comprised in a finite rule, that is to say, the human mind (even within the realm of pure mathematics) infinitely surpasses the powers of any finite machine, or else there exist absolutely unsolvable diophantine problems of the type specified.

Martin-Löf's answer based on a *constructive* interpretation of the notions of "true", "false" and "can be known" [292] is:

> There are no propositions which can neither be known to be true nor be known to be false.

For the non-constructive mathematician, the answer is:

> No propositions can be effectively produced (i.e. by an algorithm) of which it can be shown that they can neither be proved constructively nor disproved constructively. There may be absolutely unsolvable problems, but one cannot effectively produce one for which one can show that it is unsolvable.

The growing interest in this topic includes the following articles: [325, 174, 250, 127].

Finally, can we limit the effects of undecidability? One possibility is to use randomness similarly as in testing primality Section 5.9. Another possibility is to add the formal system axioms undecidable statements of the form, '$K(x) = n$', '$K(x) > n$'; see Section 2.4. Can we obtain a richer theory in this way and get some interesting consequences, still being practically sure that they are true? How to choose these statements to increase their utility? The article [46] has interesting results. Motivated by the Chaitin Theorem in Section 2.6, and one of Lipton's questions [276], Bienvenue and his colleagues probabilistically added axioms claiming the incompressibility of some strings and showed that the augmented formal systems do not prove new interesting theorems. In contrast, the axiomatic power of the statements of the form '$K(x) > n$' is significant. They showed that an augmented theory having all true statements of this type as axioms proves all true Π_1-statements.[17] In particular, their Theorem 7 shows that to derive all true Π_1-statements, it is enough to add one statement of this type for each n in case the strings x are chosen in a "special manner".

A short but insightful history of undecidability was presented in [156]. Its title, "Almost everything is undecidable", says it all! Figure 7.6, A short history of undecidability, is an illustrated summary.

Below is an English translation of the notes.

D. Hilbert dreamed of a complete system of axioms capable of establishing easily its lack of contradictions and allowing the development of entire mathematics.

In 1931, K. Gödel proved that every sufficiently powerful system for interesting and non-contradictory mathematics is incomplete: there exists true and unprovable mathematical statements in the system.

Alan Turing developed the notion of algorithm, which allowed a better understanding of the Incompleteness Theorem, and it concerns many programming problems (hence computing science).

P. Cohen proved the independence of the Axiom of Choice and the Continuum Hypothesis in the most potent set theory: undecidability influences this central mathematics theory.

Andrei Kolmogorov proposed a mathematical notion of complexity applicable to physics, hence this science is "touched" by undecidability.

Per Martin-Löf formulated rigorously, the notion of random sequence

[17]See Section 3.10.

allowing a deep understanding of the relations between probabilities and undecidability.

Yuri Matyasevich showed the undecidability of Hilbert's Tenth Problem concerning polynomial Diophantine arithmetic equations.

1. Petite histoire de l'indécidabilité

David Hilbert rêvait de trouver un système complet d'axiomes dont on pourrait établir de manière simple qu'il était non contradictoire et qui permettrait de développer toutes les mathématiques.

Kurt Gödel démontre en 1931 que tout système assez puissant pour faire des mathématiques intéressantes et non contradictoires est incomplet : certaines affirmations mathématiques vraies n'y sont pas démontrables.

Alan Turing, grâce à son travail sur la notion d'algorithmes, permet de mieux comprendre le théorème d'incomplétude et de réaliser qu'il concerne un grand nombre de problèmes de programmation (et donc d'informatique).

Paul Cohen démontre l'indépendance de l'axiome du choix et de l'hypothèse du continu au sein de la puissante théorie des ensembles : l'indécidabilité concerne donc cette théorie centrale des mathématiques.

Andrei Kolmogorov propose une définition mathématique de la notion de complexité qui s'applique en physique et qui est donc touchée par l'indécidabilité ; celle-ci est importante, y compris dans les sciences de la nature.

Per Martin-Löf, en formulant une définition rigoureuse de la notion de suite aléatoire infinie, permet une compréhension approfondie des liens entre probabilités et indécidabilité.

Yuri Matyasevich, en montrant l'indécidabilité du dixième problème de Hilbert concernant les équations polynomiales diophantiennes, permet la formulation d'équations arithmétiques – complexes – indécidables.

Jeff Paris et Leo Harrington ont formulé un énoncé d'arithmétique ne provenant pas de la logique et indécidable dans l'arithmétique élémentaire : ainsi, l'indécidabilité touche les parties les plus basiques des mathématiques.

Gregory Chaitin, en définissant le nombre Omega – dont une théorie mathématique ne peut connaître qu'un nombre fini de chiffres –, produit un concentré d'indécidabilité déconcertant.

Robert Solovay aggrave encore le résultat de G. Chaitin en produisant une variante du nombre Oméga dont *aucun* chiffre n'est connaissable par l'utilisation de la théorie usuelle des ensembles.

Raymond Smullyan tire des principes de la démonstration de Gödel un ensemble de problèmes paradoxaux qui en aident la compréhension et contribuent à populariser le résultat et sa démonstration.

Leonid Levin établit, entre autres, que l'utilisation du hasard dans les démonstrations ne permettra pas de contourner l'indécidabilité et conjecture qu'il en va de même de tout mécanisme physique.

Cristian Calude établit diverses formes fortes du théorème de Gödel qui en éclairent la nature et qui signifient que presque tout énoncé est indécidable.

© Pour la Science · n° 375 · Janvier 2009

Logique et calcul 89

Fig. 7.6: Short history of undecidability[18]

[18]Source: Jean-Paul Delahaye, email to C. Calude on 22 February 2023.

Jeff Paris and Lee Harrington formulated and proved the undecidability in the elementary arithmetic of a statement outside logic: this way, they showed that undecidability can impact the most basic mathematics.

Gregory Chaitin defined the Omega number: no mathematical theory can "know" more than finitely many of its digits – disconcerting undecidability.

Robert Solovay proved a stronger result by constructing an Omega number for which the standard set theory cannot "know" any of its digits.

Raymond Smullyan extracted principles from Gödel's proof to construct paradoxical statements, helping the understanding and popularisation of this result and its proof.

Leonid Levin showed the use of randomness in proofs will not circumvent undecidability and conjectured that this is true for physics as well.

Cristian Calude proved various strong versions of Gödel theorem, which help a better understanding of the phenomenon: they show that almost every statement is undecidable.

The Random Path

The book of days cannot be opened.
The master with the keys,
Lies away beneath the oceans,
Painting shells, deep beneath the seas.
So thus Humanity takes its course.
Sailing along selected stencils,
Gathered from random, scattered shores.[19]

Andreea S. Calude
author of
Linguistics of Social Media: An Introduction
(Routledge, 2023)

[19]In N.~Konings (ed.). *The Collection of New Zealand Poetry & Prose*, Media Publishres, Ltd., 2001, p. 45.

Postface

We have come to the end of our journey. Every page of this delightful book has another gem that contains either an insight into the halting problem or a connection between the halting problem and a different area. Cristian Calude, a world-class researcher, has done a magnificent job collecting these ideas and making them accessible to everyone.

The halting problem asks a simple question: is there a finite procedure to determine whether any program halts or continues for eternity? Martin Davis has shown us that the answer to the halting problem is negative. There does not exist any finite procedure to make such a determination. This inability is not just a technological shortcoming that will be overcome in the distant future. Rather, the halting information for all programs is inherently inaccessible. Like a black hole, we simply cannot view that information. Calude has shown that this negative result is not a small, insignificant defect in our knowledge. It does not stand alone as an isolated lacuna in our understanding of the universe. Instead, this book shows the interrelatedness of this problem with many other fields.

Why is the halting problem so central and fundamental? I believe it is important because it asks us a question that brushes up against the notion of infinity. Let me explain. Physicists inform us – with undoubted certainty – that there are 1080 particles in the universe. Thus, at every moment, our universe is finite. Our universe has only existed for a finite amount of time in the past. Within this finite universe, we are finite beings who are creating finite machines. The only aspect of our physical universe that is infinite is the potential future. Time will go on forever and ever. The halting problem asks us to determine an aspect of that infinite future. Which programs will halt, and which programs will continue forever?

Abstract mathematics always had infinities. When a mathematician considers something as simple as the set of natural numbers, they are dealing with an infinity that is associated with many unknowns. In contrast, the concrete world of computer scientists and physicists has few infinities.

Of course, one can deny the infinite future. Astrophysicists tell us of the Big Crunch (the ending counterpart to the Big Bang). Cosmologists warn us of the impending Heat Death (also called the Big Freeze or the Entropy Death) of the universe. In either scenario, no physical device will go on for eternity. We can pose a practical version of the halting problem: given a program, determine if it will halt before the Big Crunch and the looming Heat Death of the universe. This turns out to be solvable. Physicists predict when such events will occur, and we can solve the halting problem within a given time limit. However, this practical halting problem is not interesting. Many programs are known to halt but demand more time than the universe has to offer. When we avoid infinity, our universe gets less interesting.

The halting problem continues to captivate us because it deals with the physical manifestation of infinity. We are grateful to Cristian Calude for this fascinating tour of the many aspects of the halting problem.

Noson S. Yanofsky
author of
The Outer Limits of Reason:
What Science, Mathematics
and Logic Cannot *Tell Us*
(MIT Press, 2016)

Bibliography

[1] A. A. Abbott, L. Bienvenu, and G. Senno. Non-uniformity in the quantis random number generator. Report CDMTCS-472, Centre for Discrete Mathematics and Theoretical Computer Science, University of Auckland, Auckland, New Zealand, Nov. 2014.

[2] A. A. Abbott, C. S. Calude, J. Conder, and K. Svozil. Strong Kochen-Specker theorem and incomputability of quantum randomness. *Physical Review A*, 86(062109), Dec 2012.

[3] A. A. Abbott, C. S. Calude, M. J. Dinneen, and N. Huang. Experimentally probing the algorithmic randomness and incomputability of quantum randomness. *Physica Scripta*, 94:045103, 2019.

[4] A. A. Abbott, C. S. Calude, M. J. Dinneen, and N. Huang. Experimentally probing the algorithmic randomness and incomputability of quantum randomness. *Physica Scripta*, 94(4):045103, Feb 2019.

[5] A. A. Abbott, C. S. Calude, and K. Svozil. Value indefiniteness is almost everywhere. *Physical Review A*, 89(3):032109–032116, 2014.

[6] A. A. Abbott, C. S. Calude, and K. Svozil. A non-probabilistic model of relativised predictability in physics. *Information*, 6(4):773–789, 2015.

[7] A. A. Abbott, C. S. Calude, and K. Svozil. On the unpredictability of individual quantum measurement outcomes. In L. D. Beklemishev, A. Blass, N. Dershowitz, B. Finkbeiner, and W. Schulte, editors, *Fields of Logic and Computation II*, volume 9300 of *Lecture Notes in Computer Science*, pages 69–86. Springer, 2015.

[8] A. A. Abbott, C. S. Calude, and K. Svozil. A variant of the Kochen-Specker theorem localising value indefiniteness. *Journal of Mathematical Physics*, 56, 102201, `http://dx.doi.org/10.1063/1.4931658`, Oct 2015.

[9] M. Abdelhamid and G. Hassan. Blockchain and smart contracts. In *Proceedings of the 2019 8th International Conference on Software and Information Engineering*, ICSIE '19, pages 91–95, New York, NY, USA, 2019. Association for Computing Machinery.

[10] S. Abramsky. Two puzzles about computation. In S. Cooper and J. van Leeuwen, editors, *Alan Turing: His Work and Impact*, pages 53–57. Elsevier Science Publishers Ltd., 2013.

[11] W. Ackermann. Zum Hilbertschen Aufbau der reellen Zahlen. *Mathematische Annalen*, 99:118–130, 1928.

[12] L. M. Adleman. An abstract theory of computer viruses. In *Advances in Cryptology – Crypto '88*, volume 403, pages 354–374. Springer, 1988.

[13] M. Agrawal, N. Kayal, and N. Saxena. Primes is in P. *Annals of Mathematics*, 160:781–793, 2004.

[14] J. M. Agüero Trejo and C. S. Calude. A new quantum random number generator certified by value indefiniteness. *Theoretical Computer Science*, 862:3–13, Mar. 2021.

[15] J. M. Agüero Trejo and C. S. Calude. Photonic ternary quantum random number generators. *Proc. R. Soc. A*, 479:1–16, 2023.

[16] J. M. Agüero Trejo, C. S. Calude, M. J. Dinneen, A. Fedorov, A. A. Kulikov, R. Navarathna and K. Svozil. How Real Is Incomputability in Physics? Report CDMTCS-572, Centre for Discrete Mathematics and Theoretical Computer Science, University of Auckland, Auckland, New Zealand, Oct. 2023.

[17] M. Alfonseca, M. Cebrián, A. F. Anta, L. Coviello, A. Abeliuk, and I. Rahwan. Superintelligence cannot be contained: Lessons from computability theory. *Journal of Artificial Intelligence Research*, 70:65–76, 2021.

[18] E. H. Allen and C. S. Calude. Quassical computing. *International Journal of Unconventional Computing*, 14:43–57, 2018.

[19] J.-P. Allouche and J. Shallit. The ubiquitous Prouhet-Thue-Morse sequence. In T. H. C. Ding and H. Niederreiter, editors, *Sequences and their Applications. Proceedings of SETA '98*, pages 1–16. Springer, London, 1999.

[20] M. Almeida and M. Bernstein. Lucky libertarianism. *Philosophical Studies*, 113:93–119, 2003.

[21] C. Anderson. The end of theory: The data deluge makes the scientific method obsolete. *Wired Magazine*, 16.7, June 2008.

[22] G. E. Andrews. Drowning in the data deluge. *Notices Amer. Math.*

Soc., 59(7):933–941, August 2012.

[23] M. Anthony. *Discrete Mathematics of Neural Networks: Selected Topics.* SIAM, Philadelphia, PA, 2001.

[24] M. A. Aoun. Advances in three hypercomputation models. *Electronic Journal of Theoretical Physics*, 13(36):169–182, 2016.

[25] K. Appel and W. Haken. Every planar map is four-colorable, ii: Reducibility. *Illinois J. Math.*, 21:491–567, 1977.

[26] K. Appel, W. Haken, and J. Koch. Every planar map is four colourable,. *Illinois J. Math*, 21:429–490, 1977.

[27] K. I. Appel. The use of the computer in the proof of the four color theorem. *Proceedings of the American Philosophical Society*, 128(1):35–39, 1984.

[28] M. Aschbacher. The status of the classification of the finite simple groups. *Notices of the AMS*, 51:736–740, 7 2004.

[29] J. Atick and A. N. Redlich. What does the retina know about natural scenes. *Neural Computation*, (4):196–210, 1992.

[30] J. Baer, J. C. Kaufman, and R. F. Baumeister. *Are We Free? Psychology and Free Will.* Oxford University Press, New York, 2008.

[31] J. Baez. Computing the uncomputable, 2016, `johncarlosbaez.wordpress.com/2016/04/02/computing-the-uncomputable`.

[32] J. Baez. Insanely long proofs, 2012, `johncarlosbaez.wordpress.com/2012/10/19/insanely-long-proofs`.

[33] J. L. Balcázar, J. Díaz, and J. Gabarró. *Structural Complexity I, Second Edition.* Texts in Theoretical Computer Science. An EATCS Series. Springer, 1995.

[34] Y. Bar-Hillel and R. Carnap. An outline of a theory of semantic information. In Y. Bar-Hillel, editor, *Language and Information*, pages 221–274. Addison- Wesley, 1964.

[35] D. W. Barnes and J. M. Mack. *An Algebraic Introduction to Mathematical Logic.* Springer Science, 1975.

[36] N. J. Beaudry and R. Renner. An intuitive proof of the data processing inequality. *Quantum Info. Comput.*, 12(5-6):432–441, May 2012.

[37] M. J. Beeson. The mechanization of mathematics. In C. Teucher, editor, *Alan Turing: Life and Legacy of a Great Thinker*, pages 77–134. Springer, Berlin and Heidelberg, 2004.

[38] J. S. Bell. *Speakable and Unspeakable in Quantum Mechanics.* Cambridge University Press, Cambridge, 1987.

[39] S. Ben-David, P. Hrubeš, S. Moran, A. Shpilka, and A. Yehuday-

off. Learnability can be undecidable. *Nature Machine Intelligence*, 1(1):44–48, 2019.

[40] C. H. Bennett and M. Gardner. The random number Omega bids fair to hold the mysteries of the universe. *Scientific American*, 241:20–34, 1979.

[41] M. Bennett. The future of AI: What to expect in the next 5 years, `www.techtarget.com/searchenterpriseai/tip/The-future-of-AI-What-to-expect-in-the-next-5-years`, 23 May 2023.

[42] R. Bernardo-Gavito, I. E. Bagci, J. Roberts, J. Sexton, B. Astbury, H. Shokeir, T. McGrath, Y. J. Noori, C. S. Woodhead, M. Missous, U. Roedig, and R. J. Young. Extracting random numbers from quantum tunnelling through a single diode. *Scientific Reports*, 7(1):17879, 2017.

[43] C. Bernhardt. *Turing's Vision The Birth of Computer Science*. The MIT Press, Cambridge, Mass, 2016.

[44] L. Bienvenu, D. Desfontaines, and A. Shen. What percentage of programs halt? In M. M. Halldórsson, K. Iwama, N. Kobayashi, and B. Speckmann, editors, *Automata, Languages, and Programming I*, volume LNCS 9134, pages 219–230. Springer Berlin, Heidelberg, 2015.

[45] L. Bienvenu, D. Desfontaines, and A. Shen. Generic algorithms for halting problem and optimal machines revisited. *Logical Methods in Computer Science*, Volume 12, Issue 2, Apr. 2016.

[46] L. Bienvenu, A. Romashchenko, A. Shen, A. Taveneaux, and S. Vermeeren. The axiomatic power of Kolmogorov complexity. *Annals of Pure and Applied Logic*, 165(9):1380 – 1402, 2014.

[47] P. Bierhorst, E. Knill, S. Glancy, Y. Zhang, A. Mink, S. Jordan, A. Rommal, Y.-K. Liu, B. Christensen, S. W. Nam, M. J. Stevens, and L. K. Shalm. Experimentally generated randomness certified by the impossibility of superluminal signals. *Nature*, 556:223–226, April 2018.

[48] M. Bischoff. These are the most bizarre numbers in the universe. *Scientific American*, pages 1–12, `https://www.scientificamerican.com/article/these-are-the-most-bizarre-numbers-in-the-universe/`, May 2023.

[49] A. Blake and A. Zisserman. *Visual Reconstruction*. MIT Press, Cambridge, MA, 1987.

[50] W. S. Brainerd and L. H. Landweber. *Theory of Computation*. J. Wiley & Sons, New York, 1974.

[51] J. Briggs and F. D. Peat. *An Illustrated Guide to Chaos Theory and the Science of Wholeness*. Harper & Row, New York, 1989.

[52] S. Bringsjord and N. G. Sundar. In defense of the unprovability of the Church-Turing Thesis. *International Journal of Unconventional Computing*, 6:353–373, 01 2010.

[53] Č. Brukner. *Unconventional Computation. Proceedings of the 7th International Conference*, chapter Quantum Experiments Can Test Mathematical Undecidability, pages 1–5. Springer, Berlin, 2008.

[54] M. Buiatti and G. Longo. Randomness and multilevel interactions in biology. *Theory Bioscience*, 132:139–158, 2013.

[55] V. Bužek and M. Hillery. Quantum copying: Beyond the no-cloning theorem. *Phys. Rev. A*, 54:1844–1852, Sep 1996.

[56] T. Burge and H. Enderton, editors. *The Collected Works of Alonzo Church*. The MIT Press, Cambridge, MA, 2019.

[57] A. W. Burks, H. H. Goldstine, and J. von Neumann. Preliminary Discussion of the Logical Design of an Electronic Computing Instrument. 1946, 58pp.

[58] M. Burrows and D. Wheeler. A block-sorting lossless data compression algorithm. SRC Research Report 124, Systems Research Center, 1994.

[59] A. S. Calude. The journey of the four colour theorem through time. *The NZ Math. Magazine*, 38(3):27–35, 2001.

[60] C. Calude. *Theories of Computational Complexity*. North-Holland, Amsterdam, 1988.

[61] C. Calude. *Information and Randomness – An Algorithmic Perspective*. Springer, Berlin, 2002 (2nd ed.).

[62] C. Calude, H. Jürgensen, and S. Legg. Solving finitely refutable mathematical problems. In C. S. Calude, G. Păun, editors, *Finite Versus Infinite. Contributions to an Eternal Dilemma, Springer-Verlag, London*, pages 39–52, 2000.

[63] C. Calude, H. Jürgensen, and M. Zimand. Is independence an exception? *Appl. Math. Comput.*, 66:63–76, 1994.

[64] C. Calude and S. Marcus. Mathematical proofs at a crossroad? In J. Karhumäki, H. A. Maurer, G. Păun, and G. Rozenberg, editors, *Theory Is Forever, Essays Dedicated to Arto Salomaa on the Occasion of His 70th Birthday*, volume 3113 of *Lecture Notes in Computer Science*, pages 15–28. Springer, 2004.

[65] C. Calude, S. Marcus, and I. Ţevy. The first example of a recursive function which is not primitive recursive. *Hist. Math.*, 6:380–384,

1979.

[66] C. Calude and F. W. Meyerstein. Is the universe lawful? *Chaos, Solitons & Fractals*, 10(6):1075–1084, 1999.

[67] C. Calude and G. Păun. Independent instances for some undecidable problems. *RAIRO - Theoretical Informatics and Applications (RAIRO: ITA)*, 17:49–54, 1983.

[68] C. Calude and D. Vaida. Ehrenfeucht test set theorem and Hilbert basis theorem: A constructive glimpse. In *International Symposium on Mathematical Foundations of Computer Science*, pages 177–184. Springer, 1989.

[69] C. S. Calude. Chaitin Omega numbers, Solovay machines, and Gödel incompleteness. *Theor. Comput. Sci.*, 284(2):269–277, 2002.

[70] C. S. Calude. Unconventional computing: A brief subjective history. In *Advances in Unconventional Computing*, pages 855–864. Springer, 2017.

[71] C. S. Calude. Incompleteness and the Halting Problem. *Studia Logica, 1572–8730*, 2021.

[72] C. S. Calude and E. Calude. Evaluating the complexity of mathematical problems. Part 1. *Complex Systems*, 18-3:267–285, 2009.

[73] C. S. Calude and E. Calude. The complexity of the Four Colour Theorem. *LMS J. Comput. Math.*, 13:414–425, 2010.

[74] C. S. Calude and E. Calude. Evaluating the complexity of mathematical problems. Part 2. *Complex Systems*, 18-4:387–401, 2010.

[75] C. S. Calude and E. Calude. Algorithmic complexity of mathematical problems: an overview of results and open problems. *International Journal of Unconventional Computing*, 9(3-4):327–343, 2013.

[76] C. S. Calude and E. Calude. The road to quantum computational supremacy. In D. H. Bailey, N. S. Borwein, R. P. Brent, R. S. Burachik, J. H. Osborn, B. Sims, and Q. J. Zhu, editors, *From Analysis to Vizualization*, pages 249–367. Springer, 2020.

[77] C. S. Calude, E. Calude, and M. J. Dinneen. A new measure of the difficulty of problems. *Journal for Multiple-Valued Logic and Soft Computing*, 12:285–307, 2006. CDMTCS report series 277.

[78] C. S. Calude, E. Calude, and M. J. Dinneen. Adiabatic quantum computing challenges. *ACM SIGACT News*, 46(1):40–61, March 2015.

[79] C. S. Calude, E. Calude, and S. Marcus. Passages of proof. *Bull. Eur. Assoc. Theor. Comput. Sci.*, 84:167–188, 2004.

[80] C. S. Calude, E. Calude, and S. Marcus. Proving and programming. In *Randomness and Complexity, From Leibniz to Chaitin*, pages 301–

319. World Scientific, 2007.

[81] C. S. Calude, E. Calude, and M. S. Queen. The complexity of Euler's integer partition theorem. *Theor. Comput. Sci.*, 454:72–80, 2012.

[82] C. S. Calude, E. Calude, and M. S. Queen. Inductive complexity of the P versus NP problem. *Parallel Processing Letters*, 23(01):1350007, 2013.

[83] C. S. Calude, E. Calude, and K. Svozil. The complexity of proving chaoticity and the Church–Turing Thesis. *Chaos: An Interdisciplinary Journal of Nonlinear Science*, 20(3):037103, 2010.

[84] C. S. Calude, K. F. Celine, Z. Gao, S. Jain, L. Staiger, and F. Stephan. Bi-immunity over different size alphabets. *Theoretical Computer Science*, 894:31–49, 2021.

[85] C. S. Calude and D. Desfontaines. Anytime algorithms for non-ending computations. *International Journal of Foundations of Computer Science*, 26(4):465–475, 2015.

[86] C. S. Calude and D. Desfontaines. Anytime algorithms for non-ending computations. *Int. J. Found. Comput. Sci.*, 26(4):465–476, 2015.

[87] C. S. Calude and M. J. Dinneen. Exact approximations of omega numbers. *International Journal of Bifurcation and Chaos*, 17:1937–1954, 2007.

[88] C. S. Calude, M. J. Dinneen, M. Dumitrescu, and K. Svozil. Experimental evidence of quantum randomness incomputability. *Physical Review A*, 82, 022102:1–8, 2010.

[89] C. S. Calude, M. J. Dinneen, and R. Hua. QUBO formulations for the graph isomorphism problem and related problems. *Theor. Comput. Sci.*, 701:54–69, 2017.

[90] C. S. Calude, M. J. Dinneen, and C. Shu. Computing a glimpse of randomness. *Experimental Mathematics*, 11(3):361–370, 2002.

[91] C. S. Calude and M. Dumitrescu. A probabilistic anytime algorithm for the halting problem. *Computability*, 7(2-3):259–271, 2018.

[92] C. S. Calude and M. Dumitrescu. A statistical anytime algorithm for the halting problem. *Computability*, 9(2):155–166, 2020.

[93] C. S. Calude and N. J. Hay. Every computably enumerable random real is provably computably enumerable random. *Logic Journal of IGPL*, 17(4):351–374, 2009.

[94] C. S. Calude, S. Heidari, and J. Sifakis. What perceptron neural networks are (not) good for? *Inf. Sci.*, 621:844–857, 2023.

[95] C. S. Calude, S. Jain, B. Khoussainov, W. Li, and F. Stephan. Deciding parity games in quasi-polynomial time. *SIAM J. Comput.*,

0:STOC17–152–STOC17–188, 2020.

[96] C. S. Calude, F. Kroon, and N. Poznanović. Free will is compatible with randomness. *Philosophical Inquiries*, 4(2):37–52, 2016.

[97] C. S. Calude and J. P. Lewis. Is there a universal image generator? *Applied Mathematics and Computation*, 218(16):8151–8159, 2012.

[98] C. S. Calude and G. Longo. Classical, quantum and biological randomness as relative unpredictability. *Natural Computing*, 15(2):263–278, 2016.

[99] C. S. Calude and G. Longo. The deluge of spurious correlations in big data. *Foundations of Science*, 22(3):595–612, 2017.

[100] C. S. Calude, W. Meyerstein, and A. Salomaa. The universe is lawless or "pantôn chrêmatôn metron anthrôpon einai". In H. Zenil, editor, *A Computable Universe: Understanding Computation & Exploring Nature as Computation*, pages 539–547. World Scientific, Singapore, 2012.

[101] C. S. Calude and C. Müller. Formal proof: Reconciling correctness and understanding. In J. Carette, L. Dixon, C. S. Coen, and S. M. Watt, editors, *Intelligent Computer Mathematics, 16th Symposium, Calculemus 2009, 8th International Conference, MKM 2009, Held as Part of CICM 2009, Grand Bend, Canada, July 6-12, 2009. Proceedings*, volume 5625, pages 217–232, 2009.

[102] C. S. Calude and G. Păun, editors. *Finite Versus Infinite. Contributions to an Eternal Dilemma.* Spriger-Verlag, London, 2000.

[103] C. S. Calude and G. Păun. *Computing with Cells and Atoms.* Taylor & Francis Group, London and New York, 2001.

[104] C. S. Calude and L. Staiger. Long and short proofs. *Bull. Math. Soc. Sci. Math. Roumanie*, 65 (113)(2):203–211, 2022.

[105] C. S. Calude, L. Staiger, and M. Stay. *Alan Turing, His Work and Impact*, chapter Halting and non-halting Turing computations, pages 105–118. Elsevier Science Publishers, 2013.

[106] C. S. Calude and M. Stay. Natural halting probabilities, partial randomness, and zeta functions. *Inf. Comput.*, 204(11):1718–1739, 2006.

[107] C. S. Calude and M. A. Stay. Most programs stop quickly or never halt. *Advances in Applied Mathematics*, 40(3):295–308, 2008.

[108] C. S. Calude and K. Svozil. Quantum randomness and value indefiniteness. *Advanced Science Letters*, 1(1):165–168, 2008.

[109] C. S. Calude and K. Svozil. Spurious, emergent laws in number worlds. *Philosophies*, 4(2):17, 2019.

[110] C. S. Calude and D. Thompson. Incompleteness, undecidability and

automated proofs - (invited talk). In V. P. Gerdt, W. Koepf, W. M. Seiler, and E. V. Vorozhtsov, editors, *Computer Algebra in Scientific Computing – 18th International Workshop, CASC 2016, Bucharest, Romania, Proceedings*, pages 134–155, 2016.

[111] E. Calude. The complexity of Riemann's Hypothesis. *Journal for Multiple-Valued Logic and Soft Computing*, 17(4):1–9, 2011.

[112] M. Campos, S. Griffiths, R. Morris, and J. Sahasrabudhe. An exponential improvement for diagonal ramsey, 2023.

[113] G. J. Chaitin. Computational complexity and Gödel's incompleteness theorem. *ACM SIGACT News*, (9):11–12, 1971.

[114] G. J. Chaitin. A theory of program size formally identical to information theory. *Journal of the Association of Computing Machinery*, 22:329–340, 1975.

[115] G. J. Chaitin. *Algorithmic Information Theory.* Cambridge University Press, Cambridge, 1987.

[116] G. J. Chaitin. Computing the busy beaver function. In T. M. Cover and B. Gopinath, editors, *Open Problems in Communication and Computation*, page 108. Springer, New York, 1987.

[117] G. J. Chaitin. *The Unknowable.* Springer, Singapore, 1999.

[118] G. J. Chaitin. How real are real numbers? *International Journal of Bifurcation and Chaos*, 16:1841–1848, 2004.

[119] G. J. Chaitin. Leibniz, complexity and incompleteness. *APA Newsletter on Philosophy and Computers*, 9(1):7–10., 2009.

[120] G. J. Chaitin, A. Arslanov, and C. Calude. Program-size complexity computes the halting problem. *Bulletin of the EATCS*, 57, 1995, 199–200.

[121] G. J. Chaitin and J. T. Schwartz. A note on Monte Carlo primality tests and algorithmic information theory. *Communications on Pure and Applied Mathematics*, 31(4):521–527, 1978.

[122] D. J. Chalmers. *Reality+: Virtual Worlds and the Problems of Philosophy.* Penguin UK, 2022.

[123] D. G. Champernowne. The construction of decimals normal in the scale of ten. *The Journal of London Mathematical Society*, 8:254–260, 1933.

[124] L. Chen. Intuition and ingenuity: Gödel on Turing's "philosophical error". *Philosophies*, 7(2), 2022.

[125] A. Church. A note on the Entscheidungsproblem. *Journal of Symbolic Logic*, 1(1):40–41, 1936.

[126] A. Church. An unsolvable problem of elementary number theory.

American Journal of Mathematics, 58:345–363, 1936.

[127] J. Clarke-Doane. What is absolute undecidability? *Noûs*, 47(3):467–481, 2013.

[128] E. F. Codd. A relational model of data for large shared data banks. *Commun. ACM*, 13(6):77–387, 1970.

[129] F. Cohen. *Computer Viruses.* PhD thesis, University of Southern California, 1988.

[130] F. Cohen. Computational aspects of computer viruses. *Computers & Security*, 8(4):297–298, 1989.

[131] P. J. Cohen. The independence of the Continuum Hypothesis. *Proceedings of the National Academy of Sciences*, 50(6):1143–1148, 1963.

[132] M. J. Colbrook, V. Antun, and A. C. Hansen. The difficulty of computing stable and accurate neural networks: On the barriers of deep learning and Smale's 18th problem. *Proceedings of the National Academy of Sciences*, 119(12):e2107151119, 2022.

[133] W. J. Conover. *Practical Nonparametric Statistics.* John Wiley & Sons, New York, 1999.

[134] J. Conway and S. Kochen. The free will theorem. *Foundations of Physics*, 36(10):1441–1473, 2006.

[135] S. Cook. The complexity of theorem proving procedures. In *STOC '71, Proceedings of the Third Annual ACM Symposium on Theory of Computing*, pages 151–158, New York, 1971. ACM.

[136] B. S. Cooper and P. Odifreddi. Incomputability in nature. In S. B. Cooper and S. S. Goncharov, editors, *Computability and Models: Perspectives East and West*, pages 137–160. Plenum Press, New York, 2003.

[137] S. B. Cooper. Incomputability after Alan Turing. *Notices Amer. Math. Soc.*, 59(6):776–784, 2012.

[138] B. J. Copeland, editor. *The Essential Turing: Seminal Writings in Computing, Logic, Philosophy, Artificial Intelligence, and Artificial Life plus The Secrets of Enigma.* Oxford University Press, Oxford, 2004.

[139] B. J. Copeland. The Church-Turing Thesis. In E. N. Zalta, editor, *The Stanford Encyclopedia of Philosophy.* Spring 2019 edition, 2019.

[140] B. J. Copeland and O. Shagrir. The Church-Turing Thesis: Logical limit or breachable barrier? *Communications of the ACM*, 62(1):66–73, 2019.

[141] T. H. Cormen, C. E. Leiserson, R. L. Rivest, and C. Stein. *Introduction to Algorithms.* The MIT Press, 3rd edition edition, 2009.

[142] J. F. Costa. Incomputability at the foundations of physics (A study in the philosophy of science). *Journal of Logic and Computation*, 23(6):1225–1248, 2013.

[143] T. M. Cover and J. A. Thomas. *Elements of Information Theory*. John Wiley & Sons, New York, 1991.

[144] Y. Crama and P. L. Hammer. *Boolean Functions Theory, Algorithms, and Applications*. Cambridge University Press, Cambridge, England, UK, 2011.

[145] D-Wave Systems Inc. D-Wave previews next-generation quantum computing platform, `docs.dwavesys.com/docs/latest/c_gs_4.html`, 2020.

[146] N. C. A. da Costa and F. A. Doria. Undecidability and incompleteness in classical mechanics. *International Journal of Theoretical Physics*, 30:1041–1073, Aug. 1991.

[147] M. Davis. *Computability and Unsolvability*. McGraw-Hill, New York, 1958.

[148] M. Davis. *The Undecidable. Basic Papers on Undecidable, Unsolvable Problems and Computable Functions*. Raven Press, Hewlett, N.Y., 1965.

[149] M. Davis. What is a computation? In C. S. Calude, editor, *Randomness and Complexity. From Leibniz to Chaitin*, pages 89–113. World Scientific Publishing, Singapore, 2007.

[150] M. Davis. Interview with Martin Davis. *Notices Amer. Math. Soc.*, 55(560–571), May 2008.

[151] M. Davis. *The Universal Computer: the Road from Leibniz to Turing*. W. W. Norton Company, New York, 2017.

[152] M. Davis. Review of [281], *Mathematical Reviews*, MR4256190, 2023.

[153] M. Davis, Y. V. Matiyasevich, and J. Robinson. *Hilbert's tenth problem. Diophantine equations: Positive aspects of a negative solution*, pages 323–378. American Mathematical Society, Providence, RI, 1976.

[154] L. De Mol. Turing Machines. In E. N. Zalta, editor, *The Stanford Encyclopedia of Philosophy*. Metaphysics Research Lab, Stanford University, Fall 2021 edition, 2021.

[155] R. Dedekind. *Stetigkeit und irrational Zahlen*. Braunschweig, 1872 (5th ed.).

[156] J.-P. Delahaye. Presque tout est indécidable. *Pour la Science*, 375:88–93, 2009.

[157] D. C. Dennett. *Brainstorms: Philosophical Essays on Mind and Phsy-*

chology. MIT Press, 1978.

[158] N. Dershowitz and Y. Gurevich. A natural axiomatization of computability and proof of Church's Thesis. *The Bulletin of Symbolic Logic*, 14(3):299–350, 2008.

[159] D. Deutsch. Quantum theory, the Church-Turing principle and the universal quantum computer. *Proceedings of the Royal Society of London. Series A, Mathematical and Physical Sciences*, 400(1818):97–117, 1985.

[160] D. Deutsch. Quantum computation. *Physics World*, 5:57–61, 1992.

[161] P. Diaconis, S. Holmes, and R. Montgomery. Dynamical bias in the coin toss. *SIAM Review*, 49(2):211–235, 2007.

[162] E. W. Dijkstra, All Great Quotes, `www.allgreatquotes.com/quote-121574/`, 1968.

[163] M. J. Dinneen. A program-size complexity measure for mathematical problems and conjectures. In M. J. Dinneen, B. Khoussainov, and A. Nies, editors, *Computation, Physics and Beyond*, pages 81–93. Springer Berlin / Heidelberg, 2012.

[164] R. Downey and D. Hirschfeldt. *Algorithmic Randomness and Complexity*. Springer, Berlin, 2010.

[165] B. Doyle. *Free Will: The Scandal in Philosophy*. I-Phi Press, Cambridge Mass., 2011.

[166] M. du Sautoy. Deciding whether artificial intelligence can one day make great art demands a deep dive into human consciousness. *New Scientist*, 40–41, May 2019.

[167] M. Duncombe. The scandal of deduction and Aristotle's method for discovering syllogisms. *Rhizomata*, 8(2):289–311, 2020.

[168] M. I. Dyakonov. *Will We Ever Have a Quantum Computer?* Springer, Heidelberg, 2020.

[169] A. Eagle. Chance versus Randomness. In E. N. Zalta, editor, *The Stanford Encyclopedia of Philosophy*. Metaphysics Research Lab, Stanford University, Spring 2021 edition, 2021.

[170] A. S. Eddington. *The Philosophy of Physical Science*. MacMillan, 1939.

[171] A. Einstein, B. Podolsky, and N. Rosen. Can quantum-mechanical description of physical reality be considered complete? *Physical Review*, 47(10):777–780, May 1935.

[172] M. Euwe. Mengentheoretische betrachtungen über das schachspiel. *Proc. Konin. Akad. Wetenschappen (Amsterdam)*, 32(5):633–642, 1929.

[173] A. Fedorov. Realisation of the new QRNG. Personal communication to C. Calude, November 2020.

[174] S. Feferman. Are there absolutely unsolvable problems? Gödel's dichotomy. *Philosophia Mathematica*, 14(2):134–152, 2006.

[175] S. Feferman. Turing's Thesis: Ordinal logics and oracle computability. In S. B. Cooper and J. van Leeuwen, editors, *Alan Turing: His Work and Impact*, pages 144–150. Elsevier Science, Amsterdam, 2013.

[176] S. Feferman, H. M. Friedman, P. Maddy, and J. R. Steel. Does mathematics need new axioms? *Bulletin of Symbolic Logic*, 6(4):401–446, 2000.

[177] S. Feferman, J. John W. Dawson, S. C. Kleene, G. H. Moore, R. M. Solovay, and J. van Heijenoort, editors. *Kurt Gödel: Collected Works: Volume II, Publications 1938–1974*. Oxford University Press, Oxford, 1992.

[178] R. Ferber. Are correlations any guide to predictive value? *Journal of the Royal Statistical Society. Series C (Applied Statistics)*, 5(2):113–121, 1956.

[179] A. M. Ferrenberg, D. P. Landau, and Y. J. Wong. Monte Carlo simulations: Hidden errors from "good" random number generators. *Phys. Rev. Lett.*, 69:3382–3384, Dec 1992.

[180] M. Figurska, M. Stańczyk, and K. Kulesza. Humans cannot consciously generate random numbers sequences: Polemic study. *Medical Hypotheses*, 70(1):182–185, 2008.

[181] B. Fischer. *My 60 Memorable Games*. Simon & Schuster, New York, 1969.

[182] T. Franzén. *Gödel's Theorem: An Incomplete Guide to Its Use and Abuse*. AK Peter, Wellesley, Massachusetts, 2005.

[183] E. Fredkin. Digital mechanics: An informational process based on reversible universal CA. *Physica D*, 45:254–270, 1990.

[184] E. Fredkin. An introduction to digital philosophy. *International Journal of Theoretical Physics*, 42(2):189–247, 2003.

[185] S. Frieder, L. Pinchetti, R.-R. Griffiths, T. Salvatori, T. Lukasiewicz, P. C. Petersen, A. Chevalier, and J. Berner. Mathematical Capabilities of ChatGPT, `arxiv.org/abs/2301.13867`, 2023.

[186] K. E. G. E. Andrews. *Integer Partitions*. Cambridge University Press, 2004.

[187] F. Galton. Dice for statistical experiments. *Nature*, 170(42):13–14, 1890.

[188] G. Gamow. *One Two Three .... Infinity: Facts and Speculations of*

Science. Dover, New York, 1988.

[189] R. Gandy. *The Universal Turing Machine. A Half-Century Survey*, chapter The confluence of ideas in 1936. Kammerer & Unverzagt, Hamburg, 1988.

[190] N. Gisin. Indeterminism in physics, classical chaos and Bohmian mechanics: Are real numbers really real? *Erkenntnis*, 86: 1469–1481, 2021.

[191] K. Gödel. Letter to John von Neumann, `www.anilada.com/notes/godel-letter.pdf`, 1956.

[192] K. Gödel. *Über die Vollständigkeit des Logikkalküls*. PhD Thesis, University of Vienna, 1929.

[193] K. Gödel. Die vollständigkeit der axiome des logischen funktionenkalküls. *Monatshefte für Mathematik und Physik*, 37(1):349–360, 1930.

[194] K. Gödel. Über formal unentscheidbare Sätze der Principia Mathematica und verwandter Systeme I. *Monatshefte für Mathematik*, 38:173–198, 1931.

[195] K. Gödel. Über die Länge von Beweisen. *Ergebnisse eines Mathematischen Kolloquiums*, 7:23–24, 1936.

[196] K. Gödel. The consistency of the axiom of choice and of the generalized continuum-hypothesis. *Proceedings of the National Academy of Sciences*, 24(12):556–557, 1938.

[197] K. Gödel. Postscriptum. In M. Davis, editor, *The Undecidable. Basic Papers on Undecidable, Unsolvable Problems and Computable Functions*, pages 71–72. Raven Press, Hewlett, N.Y., 1965.

[198] K. Gödel. *Collected Works*. Oxford University Press, 1986.

[199] K. Gödel. Some basic theorems on the foundations of mathematics and their implications. In S. Feferman, J. W. Dawson, Jr., W. Goldfarb, C. Parsons, and R. M. Solovay, editors, *Collected Works. Unpublished Essays and Lectures. Volume III*, pages 304–323. Oxford University Press, 1995.

[200] M. Gordon. From LCF to HOL: a short history. In G. D. Plotkin, C. Stirling, and M. Tofte, editors, *Proof, Language, and Interaction, Essays in Honour of Robin Milner*, pages 169–186. The MIT Press, 2000.

[201] T. Gowers. A new proof of Szemerédi's theorem. *Geometric and Functional Analysis*, 11(3):465–588, 2001.

[202] R. Graham. Some of my favorite problems in Ramsey Theory. *INTEGERS, The Electronic Journal of Combinatorial Number Theory*,

7(2):#A2, 2007.

[203] R. Graham, B. L. Rothschild, and J. H. Spencer. *Ramsey Theory.* John Wiley and Sons, New York, 2nd edition, 1990.

[204] R. Graham and J. H. Spencer. Ramsey theory. *Scientific American,* 262:112–117, Sept. 1990.

[205] J. Grass. Reasoning about computational resource allocation. An introduction to anytime algorithms. *Magazine Crossroads*, 3(1):16–20, 1996.

[206] A. Grjebine. *La dette publique et comment s'en débarrasser.* Press Universitaire de France, Paris, 2015.

[207] A. Grünbaum. *Modern Science and Zeno's paradoxes.* Allen and Unwin, London, second edition, 1968.

[208] Y. Gurevich. Unconstrained Church-Turing Thesis cannot possibly be true. *EATCS Bulletin*, 127:45–59, February 2019.

[209] R. Guy. *Unsolved Problems in Number Theory.* Springer-Verlag, 2004.

[210] J. D. Hamkins. Every function can be computable!, `http://jdh.hamkins.org/every-function-can-be-computable/`.

[211] J. D. Hamkins and A. Miasnikov. The halting problem is decidable on a set of asymptotic probability one. *Notre Dame J. Formal Logic,* 47(4):515–524, 2006.

[212] D. Harel, A. Marron, and J. Sifakis. Autonomics: In search of a foundation for next-generation autonomous systems. *Proceedings of the National Academy of Sciences*, 117(30):17491–17498, 2020.

[213] S. Haroche and J. Raimond. Quantum computing: Dream or nightmare? *Physics Today 49, 8, 51*, 49(8):51–52, 1996.

[214] R. Harris. *The Happiness Trap: Stop Struggling, Start Living.* Trumpeter Books, Boaton, Ma., 2008.

[215] M. A. Harrison, W. L. Ruzzo, and J. D. Ullman. Protection in operating systems. *Commun. ACM*, 19(8):461–471, 1976.

[216] J. Hartmanis. On effective speed-up and long proofs of trivial theorems in formal theories. *RAIRO - Theoretical Informatics and Applications - Informatique Théorique et Applications*, 10(R1):29–38, 1976.

[217] W. Heisenberg. *Physics and Philosophy: The Revolution in Modern Science.* Harper, New York, 1958.

[218] T. Herndon, M. Ash, and R. Pollin. Does high public debt consistently stifle economic growth? A critique of Reinhart and Rogoff. *Cambridge Journal of Economics*, 38:257–279, 2014.

[219] M. Herrero-Collantes and J. C. Garcia-Escartin. Quantum random number generators. *Rev. Mod. Phys.*, 89:015004, Feb 2017.

[220] D. Hilbert. Mathematische Probleme. Vortrag, gehalten auf dem internationalen Mathematiker-Kongreßzu Paris, 1900, `https://www.deutschestextarchiv.de/book/show/hilbert_mathematische_1900`.

[221] D. Hilbert. Mathematical problems. *Bulletin of the American Mathematical Society*, 8(10):437–479, 1902.

[222] D. Hilbert. Über das Unendliche. *Mathematische Annalen*, 95(1):161–190, 1926.

[223] D. Hilbert and W. Ackermann. *Grundzüge der Theoretischen Logikg.* Springer, Berlin, 1928.

[224] L. Hiller and L. Isaacson. *Experimental Music; Composition with an Electronic Computer.* McGraw-Hill, New York – London, 1959.

[225] C. Hoefer. Causal Determinism. In E. N. Zalta and U. Nodelman, editors, *The Stanford Encyclopedia of Philosophy.* Metaphysics Research Lab, Stanford University, Spring 2023 edition, 2023.

[226] D. Hofstadter. Metafont, metamathematics, and metaphysics: Comments on Donald Knuth's article "The Concept of a Meta-Font". In *Metamagical Themas: Questing for the Essence of Mind and Pattern*, chapter 13, pages 260–296. Basic Books, New York, NY, 1985.

[227] D. R. Hofstadter. *Gödel, Escher, Bach: An Eternal Golden Braid.* Random House, New York, 1980.

[228] G. Hooft. Quantum gravity as a dissipative deterministic system. *Classical and Quantum Gravity*, 16(10):3263–3279, 1999.

[229] J. Horgan. *The End of Science. Facing the Limits of Knowledge in the Twilight of the Scientific Age.* Helix Books/Addison-Wesley, Reading, Mass., 1996.

[230] H. Huang. Induced subgraphs of hypercubes and a proof of the sensitivity conjecture. *Annals of Mathematics*, 190:949–955, 2019.

[231] M. Hunter. The Quantum Computing Era Is Here. Why It Matters. And How It May Change Our World, 2020, `tinyurl.com/tctyw8pe`.

[232] ID Quantique SA. *Quantis Certifications.* idQuantique, Geneva, Switzerland, April 2016.

[233] ID Quantique SA. *Random Number Generation – White Paper. Quantum versus Classical Random Number Generators.* idQuantique, Geneva, Switzerland, May 2020.

[234] ID Quantique SA. *Random Number Generation – White Paper. What is the Q in QRNG?* idQuantique, Geneva, Switzerland, May 2020.

[235] D. Ignatius. *The Quantum Spy.* W. W. Norton, New York, 2018.

[236] A. Jackson. Interview with Martin Davis. *Notices of the AMS*,

55(5):560–571, 2008.

[237] R. C. Johnson. World's Largest Quantum Computer Doubles Down, 2020.

[238] R. Joober and S. Karama. Randomness and nondeterminism: from genes to free will with implications for psychiatry. *Journal of Psychiatry & Neuroscience*, 20(46 (4)):doi:10.1503/jpn.210141, 2021.

[239] S. Jukna. *Boolean Function Complexity – Advances and Frontiers*, Springer, Berlin, 2012.

[240] D. Kahneman. *Thinking, Fast and Slow*. Farrar, Straus and Giroux, New York, 2011.

[241] G. Kalai. How quantum computers fail: Quantum codes, correlations in physical systems, and noise accumulation. *Quanta Magazine*, arXiv:1106.0485 [quant-ph], June 2011.

[242] G. Kalai. The argument against quantum computers, *Quanta Magazine*, www.quantamagazine.org/gil-kalais-argument-against-quantum-computers-20180207/, February 2018.

[243] L. Kalmar. An argument against the plausibility of Church's Thesis. In A. Heyting, editor, *Constructivity in Mathematics*, pages 72–80. North-Holland, Amsterdam, 1959.

[244] I. Kanter. Undecidability principle and the uncertainty principle even for classical systems. *Physical Review Letters*, 64:332–335, 1990.

[245] J. T. Kavulich, B. P. Van Deren, and M. Schlosshauer. Searching for evidence of algorithmic randomness and incomputability in the output of quantum random number generators. *Physics Letters A*, page 127032, 2020.

[246] N. L. W. Keijsers. Neural Networks. In K. Kompoliti and L. V. Metman, editors, *Encyclopedia of Movement Disorders*, pages 257–261. Academic Press, Oxford, Jan. 2010.

[247] S. C. Kleene. General recursive functions of natural numbers. *Mathematische Annalen*, 112:727–742, 1936.

[248] S. C. Kleene. *Introduction to Metamathematics*. North-Holland, Amsterdam, 1952.

[249] S. B. Kochen and E. Specker. The problem of hidden variables in quantum mechanics. *Journal of Mathematics and Mechanics (now Indiana University Mathematics Journal)*, 17(1):59–87, 1967.

[250] P. Koellner. *On the question of absolute undecidability*, pages 189–226. Lecture Notes in Logic. Cambridge University Press, 2010.

[251] B. I. Koerner. Russians engineer a brilliant slot machine cheat – and casinos have no fix. *Wired Magazine*, https://www.wired.com/20

17/02/russians-engineer-brilliant-slot-machine-cheat-casinos-no-fix/, February 2017.

[252] S. Köhler, C. Schindelhauer, and M. Ziegler. On approximating real-world halting problems. In M. Liskiewicz and R. Reischuk, editors, *Fundamentals of Computation Theory, 15th International Symposium, FCT 2005, Lübeck, Germany, August 17-20, 2005, Proceedings*, volume 3623 of *Lecture Notes in Computer Science*, pages 454–466. Springer, 2005.

[253] K. W. F. Kohlrausch. *Der experimentelle Beweis für den statistischen Charakter des radioaktiven Zerfallsgesetzes*, pages 192–212. Springer, Berlin, Heidelberg, 1926.

[254] G. Kolata. Does Gödel's Theorem matter to mathematics? In L. Harrington, M. Morley, A. Sĉedrov, and S. Simpson, editors, *Harvey Friedman's Research on the Foundations of Mathematics*, volume 117 of *Studies in Logic and the Foundations of Mathematics*, pages 399–404. Elsevier, 1985.

[255] H. Kragh. Subatomic determinism and causal models of radioactive decay, 1903-1923. RePoSS: Research Publications on Science Studies 5. Department of Science Studies, University of Aarhus, November 2009.

[256] G. Kreisel. A notion of mechanistic theory. *Synthese*, 29:11–26, 1974.

[257] S. Kritchman and R. Raz. The surprise examination paradox and the second incompleteness theorem. *Notices of the AMS*, 57(11):1454–1458, 2010.

[258] L. Kronecker. Zur theorie der elimination einer variablen aus zwei algebraischen gleichungen. *Monatsberichte der Königlich Preussischen Akademie der Wissenschaften*, pages 535–600, 1881.

[259] A. Kulikov, M. Jerger, A. Potočnik, A. Wallraff, and A. Fedorov. Realization of a quantum random generator certified with the Kochen-Specker theorem. *Phys. Rev. Lett.*, 119:240501, Dec 2017.

[260] J. C. Lagarias, editor. *The Ultimate Challenge: The 3x+1 Problem.* American Mathematical Society, 2010.

[261] T. Lampert. Wittgenstein on the infinity of primes. *History and Philosophy of Logic*, 29(1):63–81, 2008.

[262] R. Landauer. Irreversibility and heat generation in the computing process. *IBM Journal of Research and Development*, 5(3):183–191, July 1961.

[263] R. Landauer. Computation, measurement, communication and energy dissipation. In S. Haykin, editor, *Selected Topics in Signal Pro-*

cessing, page 18. Prentice Hall, Englewood Cliffs, NJ, 1989.

[264] R. Landauer. The physical nature of information. *Physics Letters A*, 217:188–0193, 1996.

[265] R. Laughlin. Nobel laureate Robert Laughlin dismisses quantum computing, www.youtube.com/watch?v=iYQSbV_BlI8, April 2020.

[266] H. D. P. Lee. *Zeno of Elea*. Cambridge University Press, Cambridge, 1936.

[267] V. Lemus, E. Acuña-Yeomans, V. Zamora, F. Hernández-Quiroz, and H. Zenil. Computable model discovery and high-level-programming approximations to algorithmic complexity. *Theoretical Computer Science*, 927:31–41, 2022.

[268] S. B. Len Sassaman, Meredith L. Patterson and A. Shubina. The Halting Problems of Network Stack Insecurity, http://langsec.org/papers/Sassaman.pdf, 2022.

[269] A. K. Lenstra, J. P. Hughes, M. Augier, J. W. Bos, T. Kleinjung, and C. Wachter. Ron was wrong, Whit is right. Santa Barbara: IACR: 17, eprint.iacr.org/2012/064.pdf, 2012.

[270] T. Lethen. Kurt Gödel's anticipation of the Turing machine: A vitalistic approach. *History and Philosophy of Logic*, 41(3):252–264, 2020.

[271] L. Levin. Universal'nye perebornye zadachi (universal search problems). *Problems of Information Transmission (Problemy Peredachi Informatsii)*, 9(3):115–116. (in Russian), 1973.

[272] C. I. Lewis and C. H. Langford. *Symbolic Logic*. Dover Publications, 1959.

[273] J. P. Lewis. Formally unsolvable problems in software engineering, citeseerx.ist.psu.edu/viewdoc/download;jsessionid=92225F617AABBB855EE871C7B0736290?doi=10.1.1.33.5250&rep=rep1&type=pdf, 1997.

[274] M. Li and P. M. B. Vitányi. *An Introduction to Kolmogorov Complexity and Its Applications*. Springer Verlag, New York, NY, 4 edition, 2019.

[275] H. W. Lin, M. Tegmark, and D. Rolnick. Why does deep and cheap learning work so well? *Journal of Statistical Physics*, 168(6):1223–1247, 2017.

[276] R. J. Lipton and K. W. Regan. Random axioms and Gödel incompleteness. https://rjlipton.wpcomstaging.com/2011/03/30/random-axioms-and-gdel-incompleteness/ Random axioms and Gödel incompleteness. rjlipton.wpcomstaging.com/2011/03/30/random-axi

`oms-and-gdel-incompleteness/`, 30 March 2011.

[277] G. Longo. Incomputability in physics. In F. Ferreira, B. Löwe, E. Mayordomo, and L. Mendes Gomes, editors, *Programs, Proofs, Processes*, pages 276–285, Berlin, Heidelberg, 2010. Springer Berlin Heidelberg.

[278] G. Longo. Incomputability in physics and biology. *Mathematical. Structures in Comp. Sci.*, 22(5):880–900, 2012.

[279] G. Longo. Interfaces of incompleteness. In G. Minati, M. R., and A. E. Pessa, editors, *Systemics of Incompleteness and Quasi-Systems*, pages 3–55. Springer Nature, June 2019.

[280] G. Longo and M. Montévil. Models and simulations: a comparison by their theoretical symmetries. In M. Dorato, L. Magnani, and T. Bertolotti, editors, *Springer Handbook of Model-Based Science*, pages 119–132. Springer, Heidelberg, 2017.

[281] S. Lucas. The origins of the Halting Problem. *Journal of Logical and Algebraic Methods in Programming*, 121:100687, 2021.

[282] J. Łukasiewicz. The shortest axiom of the implicational calculus of propositions. *Proceedings of the Royal Irish Academy*, 52(3):25–33, 1948.

[283] C. K. Lung. *Mathematical Structures of Quantum Mechanics.* World Scientific Publishing Company, Oct. 2011.

[284] R. Lupacchini. Church's Thesis, Turing's limits, and Deutsch's Principle. In M. E. Cuffaro and S. C. Fletcher, editors, *Physical Perspectives on Computation, Computational Perspectives on Physics*, pages 60–80. Cambridge University Press, Cambridge, England, UK, 2018.

[285] N. Lynch. Approximations to the Halting Problem. *Journal of Computer and System Sciences*, 9(2):143 – 150, 1974.

[286] J. Lyons. George Osborne's favourite "godfathers of austerity" economists admit to making error in research. `http://www.mirror.co.uk/news/uk-news/george-osbornes-favourite-economists-reinhart-1838219`, April 2013.

[287] B. B. Mandelbrot. A multifractal walk down wall street. *Scientific American*, 280:70–73, 1999.

[288] Y. I. Manin. Renormalisation and computation II: time cut-off and the Halting Problem. *Mathematical Structures in Computer Science*, (22):729–751, 2012.

[289] M. Mann, O. P. Sangwan, P. Tomar, and S. Singh. Automatic goal-oriented test data generation using a genetic algorithm and simulated annealing. In *2016 6th International Conference - Cloud System and*

Big Data Engineering (Confluence), pages 83–87, 2016.

[290] J. Markoff. Flaw found in an online encryption method, *New York Times*, `tinyurl.com/2n3pkzex`, 14 February 2021.

[291] P. Martin-Löf. The definition of random sequences. *Information and Control*, 9:602–619, 1966.

[292] P. Martin-Löf. Verificationism then and now. In W. D. Pauli-Schimanovich, E. Koehler, and F. Stadler, editors, *The Foundational Debate, Complexity and Constructivity in Mathematics and Physics*, pages 187–196. Kluwer Academic Publishers Dordrecht, 1995.

[293] Y. V. Matiyasevich. *Hilbert's Tenth Problem.* MIT Press, Cambridge, Massachusetts, 1993.

[294] R. McNaughton. Infinite games played on finite graphs. *Annals of Pure and Applied Logic*, 65(2):149–184, 1993.

[295] A. R. Mele. *Free Will and Luck.* Oxford University Press, Oxford, 2006.

[296] E. Mendelson. Second thoughts about Church's Thesis and mathematical proofs. *The Journal of Philosophy*, 87(5):225–233, 1990.

[297] N. Merhav. Data processing theorems and the second law of thermodynamics. *CoRR*, abs/1007.2827, 2010.

[298] D. N. Mermin. The best version of Bell's theorem. *Annals of the New York Academy of Sciences*, 755:616–623, 1995.

[299] D. N. Mermin. *Quantum Computer Science.* Cambridge University Press, Cambridge, 2007.

[300] E. Merzbacher. *Quantum Mechanics.* Wiley & Sons, New York, 3rd ed. edition, 1998.

[301] G. Miller. Riemann's hypothesis and tests for primality. *J. Comp. Syst. Sci.*, 13:300–317, 1976.

[302] M. L. Minsky. *Computation: Finite and Infinite Machines.* Prentice Hall, Inc., Englewood Cliffs, New Jersey, 1967.

[303] C. Montelle. *Chasing Shadows: Mathematics, Astronomy, and the Early History of Eclipse Reckoning.* Johns Hopkins University Press, Baltimore, 2011.

[304] R. Murawski. Undefinability of truth. The problem of the priority: Tarski vs. Gödel. *History and Philosophy of Logic*, 19:153–160, 1998.

[305] E. Nagel and J. R. Newman. *Gödel's Proof.* New York University Press, New York, 1958, 2001 (revised edition).

[306] J. Needham. *Science and Civilisation in China: Medicine*, volume 6. Cambridge University Press, 2008.

[307] E. Nelson. Internal set theory: A new approach to nonstandard anal-

ysis. *Bulletin of the American Mathematical Society*, 83(6):1165 – 1198, 1977.

[308] P. Norvig. All we want are the facts, ma'am. `http://norvig.com/fact-check.html`, 2008.

[309] F. H. Norwood. Long proofs. *The American Mathematical Monthly*, 89(2):110–112, 1982.

[310] R. Nozick. *Philosophical Explanations.* Clarendon Press, Oxford, 1981.

[311] P. Odifreddi. *Classical Recursion Theory, Vol. 1.* North-Holland, Amsterdam, 1989.

[312] C. O'Grady. Louder vowels won't get you laid, and other tales of spurious correlation. *Ars Technica*, UK, `arstechnica.co.uk/science/2015/06/louder-vowels-wont-get-you-laid-and-other-tales-of-spurious-correlation/`, 9 June 2015.

[313] M. Park, E. Leahey, and R. J. Funk. Papers and patents are becoming less disruptive over time. *Nature*, 613(7942):138–144, 2023.

[314] G. Peano. *Arithmetices Principi. Nova Methodo Exposita.* Turin Bocca, 1889.

[315] G. Peano. Sul concetto di numero. *Rivista di Matematica*, 1:87–102, 256–267, 1891.

[316] Z. Peng. Multilayer Perceptron Algebra, `arxiv.org/abs/1701.04968`, January 2017.

[317] R. Pèter. Über den zussammenhang der verschiedenen begriffe der rekursiven funktion. *Mathematische Annalen*, 110:612–632, 1934.

[318] M. Petrov, I. Radchenko, D. Steiger, R. Renner, M. Troyer, and V. Makarov. Independent quality assessment of a commercial quantum random number generator. *EPJ Quantum Technology*, 9(1):17, 2022.

[319] N. Pippenger. The shortest disjunctive normal form of a random boolean function. *Random Structures & Algorithms*, 22(2):161–186, 2003.

[320] S. Pironio. The certainty of randomness. *Nature*, 556:176–177, 2018.

[321] J. Poppelaars. OR at Work, `https://john-poppelaars.blogspot.com`, April 2015.

[322] K. R. Popper. *Objective Knowledge: An Evolutionary Approach.* Oxford University Press, 1972.

[323] E. L. Post. Introduction to a general theory of elementary propositions. *American Journal of Mathematics*, 43(3):163–185, 1921.

[324] E. L. Post. Finite combinatory processes – formulation 1. *Journal of*

Symbolic Logic, 1:103–105, 1936.

[325] E. L. Post. Formal reductions of the general combinatorial decision problem. *American Journal of Mathematics*, 65:197–215, 1943.

[326] M. B. Pour-El and J. I. Richards. *Computability in Analysis and Physics*. Springer, Berlin, 1989.

[327] P. Pudlák. On the length of proofs of finitistic consistency statements in first order theories. In J. Paris, A. Wilkie, and G. Wilmers, editors, *Logic Colloquium '84*, volume 120 of *Studies in Logic and the Foundations of Mathematics*, pages 165–196. Elsevier, 1986.

[328] P. Pudlák. *Handbook Proof Theory*, chapter The lenghts of proofs, pages 548–637. Elsevier Science Publishers, 1998.

[329] G. K. Pullum. Scooping the loop snooper. A proof that the halting problem is undecidable, `http://www.lel.ed.ac.uk/~gpullum/loopsnoop.html`, 2008, 2022.

[330] H. Putnam. *Reason, Truth and History*. Cambridge University Press, Cambridge, 1981.

[331] M. O. Rabin. Decidability of second-order theories and automata on infinite trees. *Transactions of the American Mathematical Society*, 141:1–35, 1069.

[332] M. O. Rabin. Probabilistic algorithm for testing primality. *J. Number Th.*, 12,:128–138, 1980.

[333] T. Rado. On non-computable functions. *The Bell System Technical Journal*, XLI(41)(3):877–884, May 1962.

[334] H. Ragab, A. Milburn, K. Razavi, H. Bos, and C. Giuffrida. Crosstalk: Speculative data leaks across cores are real. In *2021 IEEE Symposium on Security and Privacy (SP)*, pages 1852–1867, 2021.

[335] M. Reck, A. Zeilinger, H. J. Bernstein, and P. Bertani. Experimental realization of any discrete unitary operator. *Physical Review Letters*, 73:58–61, 1994.

[336] H. Reichenbach. *The Theory of Probability*. University of California Press, 1949.

[337] C. Reinhart and K. Rogoff. Growth in a time of debt. *American Economic Review*, 2:573–578, 2010.

[338] G. Restall. Paraconsistency everywhere. *Notre Dame Journal of Formal Logic*, 43(3):147–156, 2002.

[339] H. G. Rice. Classes of recursively enumerable sets and their decision problems. *Transactions of the American Mathematical Society*, 74(2):358–366, 1951.

[340] H. Riesel. *Prime Numbers and Computer Methods for Factoriza-*

tion, chapter Classical Methods of Factorization, pages 141–172. Birkhäuser Boston, MA, 2012.

[341] N. Robertson, D. P. Sanders, P. Seymour, and R. Thomas. The four-colour theorem. *J. Combinatorial Theory, Series B*, 70:2–44, 1997.

[342] H. Rogers, Jr. *Theory of Recursive Functions and Effective Computability*. MacGraw-Hill, New York, 1967.

[343] R. Rojas. *Neural Networks*. Springer, Berlin, 1996.

[344] R. Roscher, B. Bohn, M. F. Duarte, and J. Garcke. Explainable machine learning for scientific insights and discoveries. *IEEE Access*, 8:42200–42216, 2020.

[345] J. Rosen. *Lawless Universe: Science and the Hunt for Reality*. The John Hopkins University Press, Baltimore, Maryland, 2010.

[346] S. D. Ross. *Perspective in Whitehead's Metaphysics*. State University of New York, Albany, USA, 1983.

[347] M. Route. Intel secure key-powered radio-flaring ultracool dwarf population synthesis. In *American Astronomical Society Meeting Abstracts #234*, volume 234 of *American Astronomical Society Meeting Abstracts*, id. 207.01, 2019.

[348] D. Ruderman and B. Bialek. Statistics of natural images: scaling in the woods. *Physical Review Letters*, (73):814–817, 1994.

[349] B. Russell. *Our Knowledge of the External World*. Routledge, London, 2009.

[350] B. Russell and A. N. Whitehead. *Principia Mathematica*. Cambridge University Press (three volumes), 1910, 1912, 1913.

[351] A. Rybalov. On the strongly generic undecidability of the halting problem. *Theoretical Computer Science*, 377(1):268–270, 2007.

[352] S. Sehra, D. Flores, and G. D. Montañez. Undecidability of underfitting in learning algorithms. In *2021 2nd International Conference on Computing and Data Science (CDS)*, pages 591–594, 2021.

[353] A. A. Selçuk, F. Orhan, and B. Batur. Undecidable problems in malware analysis. In *2017 12th International Conference for Internet Technology and Secured Transactions (ICITST)*, pages 494–497, 2017.

[354] J. Sessions. The Case For Growth: Sessions Lists Benefits Of Discretionary Cuts, 2011, `www.sessions.senate.gov/public/index.cfm/news-releases?ID=E36C43B4-B428-41A4-A562-475FC16D3793`, March 2011.

[355] C. E. Shannon. A mathematical theory of communication. *Bell System Tech. J.*, 27(4):623–666, 1948. Reprinted in C. E. Shannon and W. Weaver, *The Mathematical Theory of Communication*, University

of Illinois Press, Urbana, Illinois, 1949.

[356] S. Shapiro. Incompleteness and inconsistency. *Mind*, 111(444):817–832, 10 2002.

[357] S. J. Shapiro. *Fancy Bear Goes Phishing: The Dark History of the Information Age, in Five Extraordinary Hacks.* Penguin Random House, 2023.

[358] I. Shumailov, Z. Shumaylov, Y. Zhao, Y. Gal, N. Papernot, and R. Anderson. The curse of recursion: Training on generated data makes models forget, `news.ycombinator.com/item?id=36319076`, 2023.

[359] J. Sifakis. Can we trust autonomous systems? Boundaries and risks. In Y. Chen, C. Cheng, and J. Esparza, editors, *Automated Technology for Verification and Analysis - 17th International Symposium, ATVA 2019, Taipei, Taiwan, October 28-31, 2019, Proceedings*, volume 11781 of *Lecture Notes in Computer Science*, pages 65–78. Springer, 2019.

[360] J. Sifakis. Testing system intelligence, `https://doi.org/10.48550/arXiv.2305.11472`, 2023.

[361] J. Sifakis and D. Harel. Trustworthy autonomous system development. *ACM Trans. Embed. Comput. Syst.*, 22(3):40:1–40:24, 2023.

[362] M. Sipser. *Introduction to the Theory of Computation.* International Thomson Publishing, 2013 (3rd ed.).

[363] A. Skodras, C. Christopoulos, and T. Ebrahimi. The JPEG 2000 still image compression standard. *IEEE Signal Processing Magazine*, 18(5):36–58, 2001.

[364] J. Smith. From Reinhart & Rogoff's own data: UK GDP increased fastest when debt-to-GDP ratio was highest – and the debt ratio came down!, `tinyurl.com/yj5u76s2`, April 2013.

[365] P. Smith. *Explaining Chaos.* Cambridge University Press, 1998.

[366] P. Smith. *An Introduction to Gödel's Theorems.* Cambridge University Press, Cambridge, UK, 2013.

[367] E. Snapper. The three crises in mathematics: Logicism, intuitionism and formalism. *Mathematics Magazine*, 54(4):207–216, 1979.

[368] R. I. Soare. *Handbook of Computability Theory*, chapter The History and Concept of Computability, pages 3–36. North-Holland, 1999.

[369] D. Sobania, M. Briesch, C. Hanna, and J. Petke. An analysis of the automatic bug fixing performance of Chatgpt, `arxiv.org/abs/2301.08653`, 2023.

[370] F. Soler-Toscano, H. Zenil, J.-P. Delahaye, and N. Gauvrit. Calculat-

ing Kolmogorov complexity from the output frequency distributions of small turing machines. *PLoS ONE*, 9(5):e96223, 2014.

[371] R. Solovay and V. Strassen. A fast Monte-Carlo test for primality. *SIAM Journal on Computing*, 6(1):84–85, 1977.

[372] R. Solovay and V. Strassen. Erratum: A fast Monte-Carlo test for primality. *SIAM Journal on Computing*, 7(1):118, 1978.

[373] R. M. Solovay. On random r.e. sets. In A. I. Arruda, N. C. A. da Costa, and R. Chuaqui, editors, *Non-Classical Logics, Model Theory, and Computability*, pages 283–307. North-Holland, Amsterdam, 1977.

[374] G. Sommaruga, editor. *Formal Theories of Information: From Shannon to Semantic Information Theory and General Concepts of Information*, volume 5363 of *Lecture Notes in Computer Science*. Springer, 2009.

[375] E. Specker. Nicht konstruktiv beweisbare Sätze der Analysis. *The Journal of Symbolic Logic*, 14:145–158, 1949.

[376] J. Spencer. Short theorems with long proofs. *The American Mathematical Monthly*, 90(6):365–366, 1983.

[377] L. Staiger. A tight upper bound on Kolmogorov complexity and uniformly optimal prediction. *Theory of Computing Systems*, 31(3):215–229, 1998.

[378] B. Steinbach and R. Kohut. Neural networks – a model of Boolean functions. In *Boolean Problems, Proceedings of the 5th International Workshop on Boolean Problems*, pages 223–240, 2002.

[379] T. Stephen and T. Yusun. Counting inequivalent monotone boolean functions. *Discrete Applied Mathematics*, 167:15–24, 2014.

[380] C. Stokel-Walker. Chatgpt can find and fix the bugs in computer code. *The New Scientist*, page 16, `tinyurl.com/28mhpm4d`, 4 February 2023.

[381] C. Strachey. An impossible program. *The Computer Journal*, 7(4):313–313, 1965.

[382] G. Strawson. The impossibility of moral responsibility. *Philosophical Studies*, 75:5–24, 1994.

[383] G. Sudan. Sur le nombre transfinite ω^{ω}. *Bull. Math. Soc. Roumaine des Sciences*, 30:11–30, 1927.

[384] P. Suppes. The transcendental character of determinism. *Midwest Studies In Philosophy*, 18(1):242–257, 1993.

[385] K. Svozil. The quantum coin toss – testing microphysical undecidability. *Physics Letters A*, 143:433–437, 1990.

[386] T. Tao. Embracing change and resetting expectations, `unlocked.microsoft.com/ai-anthology/terence-tao/`, 12 June 2023.

[387] T. Tao. Almost all orbits of the Collatz map attain almost bounded values. *Forum of Mathematics, Pi*, 10:e12, 2022.

[388] A. Tarski. *Pojęcie Prawdy w Językach Nauk Dedukcyjnych.* Nakładem Towarzystwa Naukowego Warszawskiego, 1933.

[389] A. Tarski. Der Wahrheitsbegriff in den Sprachen der deduktiven Disziplinen. *Akademie der Wissenschaften in Wien. Mathematisch-naturwissenschaftliche Klasse, Akademischer Anzeiger*, 69:9–12, 1936.

[390] A. Tarski. *A Decision Method for Elementary Algebra and Geometry.* University of California Press, Berkeley and Los Angeles, 1951.

[391] P. Teller. Computer proof. *Journal of Philosophy*, 77(12):797–803, 1980.

[392] P. Thagard. Hard decisions, `www.psychologytoday.com/nz/blog/hot-thought/201408/hard-decisions`, 2014.

[393] The RAND Corporation. *A Million Random Digits with 100,000 Normal Deviates Free Press Publishers.* Knolls Atomic Power Lab. Report KAPL-3147, Glencoe, Illinois, 1955.

[394] R. Thomas. John Conway – Discovering Free Will (part iii), Plus+, `plus.maths.org/content/john-conway-discovering-free-will-part-iii?src=aop`, 27 December 2011.

[395] D. Thompson. Formalisation vs. understanding. In C. S. Calude and M. J. Dinneen, editors, *Unconventional Computation and Natural Computation: 14th International Conference, UCNC 2015, Auckland, New Zealand, Proceedings*, pages 290–300. Springer, Heidelberg, 2015.

[396] T. Toffoli. Reversible computing. Technical report, MIT/LCS Tecnhnical Memo 151, February 1980, 38 pp.

[397] A. M. Turing. On computable numbers, with an application to the Entscheidungsproblem. *Proceedings of the London Mathematical Society, Series 2*, 42, 43:230–265, 544–546, 1936-7 and 1937.

[398] A. M. Turing. Systems of logic based on ordinals. *Proceedings of the London Mathematical Society, Series 2*, 45:161–228, 1939.

[399] A. M. Turing. Intelligent machinery. In C. R. Evans and A. D. J. Robertson, editors, *Cybernetics: Key Papers*, pages 27–52. Butterworths, London, 1968.

[400] T. Tymoczko. The four-color problem and its philosophical significance. *The Journal of Philosophy*, 76(2):57–83, 1979.

[401] J. van Heijenoort, editor. *From Frege to Gödel: A Source Book in*

Mathematical Logic, 1879–1931, chapter The foundations of elementary arithmetic, pages 302–333. Harvard University Press, 1967.

[402] P. van Inwagen. *An Essay on Free Will.* Clarendon Press, Oxford, 1983.

[403] M. Y. Vardi. Solving the unsolvable. *Communications of the ACM*, 54(7):5, 2011.

[404] T. Vigen. *Spurious Correlations.* Hachette Books, New York, 2015.

[405] J. von Neumann. *Theory of Self-Reproducing Automata.* University of Illinois Press, Urbana, 1966. A. W. Burks, editor.

[406] W. A. Wagenaar. Generation of random sequences by human subjects: A critical survey of literature. *Psychology Bulletin*, 77(1):65–72, 1972.

[407] E. J. Weyuker. The applicability of program schema results to programs. *International Journal of Computer & Information Sciences*, 8(5):387–403, 1979.

[408] E. P. Wigner. The unreasonable effectiveness of mathematics in the natural sciences. Richard Courant Lecture delivered at New York University, May 11, 1959. *Communications on Pure and Applied Mathematics*, 13:1–14, 1960.

[409] Wikipedia contributors. Morris worm – Wikipedia, the free encyclopedia. [Online; accessed 9-June-2022], `en.wikipedia.org/w/index.php?title=Morris_worm&oldid=1083821449`, 2022.

[410] R. Wilson. *Four Colours Suffice.* Penguin, London, 2002.

[411] S. Wolfram. Computational foundations for the Second Law of Thermodynamics.`writings.stephenwolfram.com/2023/02/computational-foundations-for-the-second-law-of-thermodynamics/`, February 2023.

[412] S. Wolfram. The Wolfram Physics Project: A One-Year Update.`writings.stephenwolfram.com/2021/04/the-wolfram-physics-project-a-one-year-update/`, April 2021.

[413] S. Wolfram. What Is Chatgpt Doing ... and Why Does It Work?, `writings.stephenwolfram.com/2023/02/what-is-chatgpt-doing-and-why-does-it-work/`, March 2023.

[414] S. Wolfram. Undecidability and intractability in theoretical physics. *Physical Review Letters*, 54(8):735–738, Feb 1985.

[415] S. Wolfram. *A New Kind of Science.* Wolfram Media, Inc., Champaign, IL, 2002.

[416] A. Wright. Is it even possible to be "completely secure"?, Medium, `medium.com/@aprilcwright/is-it-even-possible-to-be-compl`

etely-secure-6c7a92a297a9, 9 january 2019.

[417] J. Xu, X. Zhang, and C. Urban. Mechanising Turing Machines and Computability Theory in Isabelle/Hol. In S. Blazy, C. Paulin-Mohring, and D. Pichardie, editors, *Interactive Theorem Proving*, pages 147–162, Berlin, Heidelberg, 2013. Springer Berlin Heidelberg.

[418] N. Yanofsky. Paradoxes, contradictions, and the limits of science. *American Scientist*, 104:166–173, 2016.

[419] A. Young and M. Yung. Cryptovirology: extortion-based security threats and countermeasures. In *Proceedings 1996 IEEE Symposium on Security and Privacy*, pages 129–140, 1996.

[420] S. Yu-Ting. A "natural" enumeration of non-negative rational numbers – an informal discussion. *The American Mathematical Monthly*, (1):25–29.

[421] A. Zeilinger. The message of the quantum. *Nature*, 438:743, 2005.

[422] A. Zeilinger. Spooky action and beyond, *Signandsight*, www.signandsight.com/features/614.html, 16 February, 2006.

[423] H. Zenil. Computer runtimes and the length of proofs. In M. J. Dinneen, B. Khoussainov, and A. Nies, editors, *Computation, Physics and Beyond*, volume LNCS 7160, pages 224–240. Springer Berlin, Heidelberg, 2012.

[424] H. Zenil, editor. *Computable Universe: Understanding and Exploring Nature as Computation.* World Scientific, Singapore, 2013.

[425] H. Zenil and J.-P. Delahaye. On the algorithmic nature of the world. In G. Dodig-Crnkovic and M. Burgin, editors, *Information and Computation. Essays on Scientific and Philosophical Understanding of Foundations of Information and Computation*, pages 477–499. World Scientific, 2010.

[426] W. Zielonka. Infinite games on finitely coloured graphs with applications to automata on infinite trees. *Theoretical Computer Science*, 200:135–183, 1998.

[427] J. Ziv and A. Lempel. Compression of individual sequences via variable-rate coding. *IEEE Transactions on Information Theory*, 24:530–536, 1978.

[428] K. Zuse. *Rechnender Raum.* Friedrich Vieweg & Sohn, Braunschweig, 1969.

[429] Apple rushes to block 'zero-click' iPhone spyware, www.bbc.com/news/business-58540936, September 2021.

[430] The flyspeck project, en.wikipedia.org/wiki/Kepler_conjecture#A_formal_proof, 10 August 2014.

[431] Halting problem, codedocs.org/what-is/halting-problem, 2023.

[432] Isabelle 2022, isabelle.in.tum.de, 2023.

[433] List of long mathematical proofs – Wikipedia, The Free Encyclopedia. en.wikipedia.org/w/index.php?title=List_of_long_mathematical_proofs&oldid=1029115674, 2021. [Online; accessed 18-May-2022].

[434] Stigler's law of eponymy, en.wikipedia.org/wiki/Stigler%27s_law_of_eponymy, 2023.

[435] Turing's proof, en.wikipedia.org/w/index.php?title=Turing%27s_proof&oldid=1020372460.

Index

www.ingramcontent.com/pod-product-compliance
Lightning Source LLC
LaVergne TN
LVHW010331220926
848484LV00004BB/882